Innovation in Propagation of Fruit, Vegetable and Ornamental Plants

Innovation in Propagation of Fruit, Vegetable and Ornamental Plants

Editor

Sergio Ruffo Roberto

MDPI • Basel • Beijing • Wuhan • Barcelona • Belgrade • Manchester • Tokyo • Cluj • Tianjin

Editor
Sergio Ruffo Roberto
Londrina State University,
Brazil

Editorial Office
MDPI
St. Alban-Anlage 66
4052 Basel, Switzerland

This is a reprint of articles from the Special Issue published online in the open access journal *Horticulturae* (ISSN 2311-7524) (available at: https://www.mdpi.com/journal/horticulturae/special_issues/innovation_propagation).

For citation purposes, cite each article independently as indicated on the article page online and as indicated below:

LastName, A.A.; LastName, B.B.; LastName, C.C. Article Title. *Journal Name* **Year**, *Article Number, Page Range*.

ISBN 978-3-03943-410-7 (Hbk)
ISBN 978-3-03943-411-4 (PDF)

Cover image courtesy of Sergio R. Roberto.

Contents

About the Editor . . . . **vii**

Sergio Ruffo Roberto and Ronan Carlos Colombo
Innovation in Propagation of Fruit, Vegetable and Ornamental Plants
Reprinted from: *Horticulturae* 2020, 6, 23, doi:10.3390/horticulturae6020023 . . . . **1**

Chaokun Huang, Yuexia Wang, Yanjuan Yang, Chuan Zhong, Michitaka Notaguchi and Wenjin Yu
A Susceptible Scion Reduces Rootstock Tolerance to *Ralstonia solanacearum* in Grafted Eggplant
Reprinted from: *Horticulturae* 2019, 5, 78, doi:10.3390/horticulturae5040078 . . . . **9**

**Renata Koyama, Walter Aparecido Ribeiro Júnior, Douglas Mariani Zeffa,
Ricardo Tadeu Faria, Henrique Mitsuharu Saito, Leandro Simões Azeredo Gonçalves and
Sergio Ruffo Roberto**
Association of Indolebutyric Acid with *Azospirillum brasilense* in the Rooting of Herbaceous Blueberry Cuttings
Reprinted from: *Horticulturae* 2019, 5, 68, doi:10.3390/horticulturae5040068 . . . . **25**

Alexis Galus, Ali Chenari Bouket and Lassaad Belbahri
In Vitro Propagation and Acclimatization of Dragon Tree (*Dracaena draco*)
Reprinted from: *Horticulturae* 2019, 5, 64, doi:10.3390/horticulturae5030064 . . . . **31**

Leo Sabatino, Fabio D'Anna and Giovanni Iapichino
Improved Propagation and Growing Techniques for Oleander Nursery Production
Reprinted from: *Horticulturae* 2019, 5, 55, doi:10.3390/horticulturae5030055 . . . . **39**

Hasan Mehraj, Md. Meskatul Alam, Sultana Umma Habiba and Hasan Mehbub
LEDs Combined with CHO Sources and CCC Priming PLB Regeneration of *Phalaenopsis*
Reprinted from: *Horticulturae* 2019, 5, 34, doi:10.3390/horticulturae5020034 . . . . **49**

**Angel R. Del Valle-Echevarria, Michael B. Kantar, Julianne Branca, Sarah Moore,
Matthew K. Frederiksen, Landon Hagen, Tanveer Hussain and David J. Baumler**
Aeroponic Cloning of *Capsicum* spp.
Reprinted from: *Horticulturae* 2019, 5, 30, doi:10.3390/horticulturae5020030 . . . . **61**

Thomas Gradziel, Bruce Lampinen and John E. Preece
Propagation from Basal Epicormic Meristems Remediates an Aging-Related Disorder in Almond Clones
Reprinted from: *Horticulturae* 2019, 5, 28, doi:10.3390/horticulturae5020028 . . . . **69**

Carolina Smanhotto Schuchovski and Luiz Antonio Biasi
In Vitro Establishment of 'Delite' Rabbiteye Blueberry Microshoots
Reprinted from: *Horticulturae* 2019, 5, 24, doi:10.3390/horticulturae5010024 . . . . **79**

Tales Poletto, Valdir Marcos Stefenon, Igor Poletto and Marlove Fátima Brião Muniz
Pecan Propagation: Seed Mass as a Reliable Tool for Seed Selection
Reprinted from: *Horticulturae* 2018, 4, 26, doi:10.3390/horticulturae4030026 . . . . **93**

About the Editor

Sergio Ruffo Roberto is a graduate of Agronomy, and has a master's degree and PhD in Plant Production from Universidade Estadual Paulista—UNESP, Brazil, and a Post-Doctorate qualification from the U.S. Horticultural Research Laboratory—Agricultural Research Service, USDA, United States. Since 2000, Dr. Roberto has been an Associate Professor in the area of Fruit Crops at the Universidade Estadual de Londrina, Brazil, working in the areas of Management, Propagation and the Physiology of Temperate Fruit Crops. Professor Roberto offers a course on Fruit Crops in the Agronomy Undergraduate Course, and Fruit Production Technology in the Plant Production Graduate Program.

 *horticulturae*

Editorial

Innovation in Propagation of Fruit, Vegetable and Ornamental Plants

Sergio Ruffo Roberto [1,*] **and Ronan Carlos Colombo** [2]

[1] Agricultural Research Center, Londrina State University, Celso Garcia Cid Road, km 380, P.O. Box 10.011, Londrina 86057-970, Brazil

[2] Agricultural Science Department, Federal Technological University of Parana, Francisco Beltrão 85601-970, Brazil; ronancolombo@utfpr.edu.br

* Correspondence: sroberto@uel.br; Tel.: +55-43-3371-4774

Received: 22 January 2020; Accepted: 30 January 2020; Published: 9 April 2020

Abstract: There are two primary forms of plant propagation: sexual and asexual. In nature, propagation of plants most often involves sexual reproduction, and this form is still used in several species. Over the years, horticulturists have developed asexual propagation methods that use vegetative plant parts. Innovation in plant propagation has supported breeding programs and allowed the production of high-quality nursery plants with the same genetic characteristics of the mother plant, and free of diseases or pests. The purpose of this Special Issue, "Innovation in Propagation of Fruit, Vegetable and Ornamental Plants", was to present state-of-the-art techniques recently developed by researchers worldwide. The Special Issue has brought together some of the latest research results of new techniques in plant propagation in nine original papers, which deal with a wide range of research activities.

Keywords: nursery plants; plant multiplication; seeds; cuttings; budding; grafting; micropropagation; biotechnology

1. Introduction

In horticulture, plant propagation plays an important role as the number of plants can be rapidly multiplied retaining desirable characteristics of the mother plants, as well as reducing the bearing age of the plants. Depending on the species, different techniques can be applied to optimize a nursery production system or even to solve a specific propagation difficulty.

In cultivated almond (*Prunus dulcis*), non-infectious bud-failure (NBF) is a commercially important age-related disorder that results in the failure of new vegetative buds to grow in the spring. The incidence of NBF increases with clone age, including within individual long-lived trees as well as nursery propagation lineages. Consequently, nursery practices emphasize the establishment of foundation-mother blocks, utilizing propagation-wood selected from proven and well-monitored propagation lineages. Commercial propagation utilizes axillary shoot buds through traditional budding or grafting. Thus, to solve this issue, one possibility is to examine NBF development using basal epicormic buds from individual trees of advanced age as an alternative source of foundation stock.

Blueberry (*Vaccinium virgatum*) traditionally is propagated by softwood, semi-hardwood and hardwood cuttings or even rhizome cuttings of selected clones [1]. Some challenges in this production are a very low rooting percentage in many genotypes, the amount of time required to propagate and commercialize newly-released cultivars for mass propagation [1,2] and phytosanitary problems. In vitro culture (micropropagation) can overcome the limitations of traditional cuttings, presenting an alternative for faster growth throughout the year (with no seasonal effects) without pathogens.

Seed production in pepper (*Capsicum* sp.) is achieved by the self-pollination of promising plants to develop true breeding lines. In recent years, there has been an increase in the use of hybrid cultivars because of their higher yields and often unique characteristics. However, seeds from hybrid plants cannot be saved for the next generation since self-progeny would show segregation of desirable traits. Therefore, a clonal propagation protocol could be implemented to enhance profitability of these novel germplasm types [3].

Bacterial wilt is a soil-borne disease of eggplant (*Solanum melongena*) caused by *Ralstonia solanacearum*, a highly pathogenic soil-borne bacterium that invades the vascular system of a host plant leading to plant wilting and death. At present, the use of tolerant rootstocks for grafting eggplant varieties is the most effective approach to control bacterial wilt disease [4]. Thus, different grafted combinations of eggplant need to be studied to determine if they affect plant growth and abate the bacterial wilt tolerance of a rootstock upon inoculation with *R. solanacearum*.

The Dragon Tree or Drago (*Dracaena draco*) is a subtropical ornamental plant used for its medicinal properties. In view of the drastic reduction in the number of individual varieties and the consequent loss of dragon tree genetic diversity caused by absence of natural regeneration, micropropagation offers many advantages because it potentially can facilitate large-scale production of valuable clones and allow plant reintroduction in its natural ecosystem [5–9]. Unfortunately, most of the published protocols have been poorly described, in particular concerning the difficult stage of acclimatization, so it is necessary to develop an efficient procedure for micropropagation and subsequent acclimatization of this species under in vitro culture conditions.

2. Papers in This Special Issue

The Special Issue "Innovation in Propagation of Fruit, Vegetable and Ornamental Plants" brings together some of the latest research results of new techniques in this field. It presents nine original papers, which deal with a wide range of research activities.

We can divide the Special Issue in three parts, as follows.

2.1. Fruit Crops

The first contribution in this section explored the "Association of Indolebutyric Acid with *Azospirillum brasilense* in the Rooting of Herbaceous Blueberry Cuttings" by Koyama et al. [10]. The use of plant growth-promoting rhizobacteria (PGPRs) can be a promising biological alternative for increasing the rooting of blueberry (*Vaccinium* sp.) cuttings [11]. *Azospirillum* is a genus of PGPRs that inhabits the roots of host plants and provides beneficial effects to the plant under normal growth and/or stress conditions [12]. PGPRs of this genus can increase the fixation of free nitrogen and the production of phytohormones, thus promoting growth in inoculated plants [13].

Studies have shown promising results for association with the PGPR species *Azospirillum brasilense*, since it promotes root development by increasing the production of hormones, leading to growth and development of plants [14–16]. In this context, the objective of this study was to assess the viability of producing blueberry nursery plants from cuttings using different doses of IBA in association with *A. brasilense*. The authors found that the application of IBA with the *A. brasilense* rhizobacteria increased the number of roots of 'Powderblue' blueberry cuttings, while the treatments with indolebutyric acid (IBA) alone and IBA 1000 mg L^{-1} + *A. brasilense* increased the root length of the cuttings. However, treatments with IBA and *A. brasilense* combined had no impact on the % rooted cuttings and % survival of the cuttings.

The second paper illustrated that "Propagation from Basal Epicormic Meristems Remediates an Aging-Related Disorder in Almond Clones" by Gradziel et al. [17]. The deterioration of clone performance with age has similarly been reported in several crops [18–20] due to genetic [20] or epigenetic [21,22] changes resulting in losses in productivity and/or crop value. Non-infectious bud-failure (NBF) is a disorder of almond (*Prunus dulcis*) characterized by the failure of terminal vegetative buds to push in the spring [23]. Extensive research has failed to find any association of NBF with known pathogens, including viruses and viroids [24–28]. However, the incidence of NBF increases with clone age, including within individual long-lived trees as well as nursery propagation lineages. Thus, the aim of this study was to examine NBF development using basal epicormic buds from individual trees of advanced age as an alternative source of foundation stock, since almond commercial propagation utilizes axillary shoot buds through traditional budding or grafting.

The results of this study showed that the age-related progression of NBF was suppressed in these epicormic meristems, possibly owing to their unique origins and ontogeny. However, the underlying mode of action for suppression remains unknown, and vegetative-progeny testing remains the only effective strategy for identifying suitable commercial propagation clone sources. NBF development in commercial orchards propagated from foundation blocks established from these sources was similarly dramatically suppressed even over the 10- to 20-year expected commercial orchard life. Foundation-stock stability can be further maintained through appropriate management of propagation source trees, which requires accurate knowledge of meristem origin and development.

The third paper concerned the "In Vitro Establishment of 'Delite' Rabbiteye Blueberry Microshoots" by Schuchovski and Biasi [29]. Traditionally, blueberry is propagated by softwood, semi-hardwood and hardwood cuttings [1], or even rhizome cuttings of selected clones [30]. Some challenges in this production are a very low rooting percentage in many genotypes, the amount of time required to propagate and commercialize newly-released cultivars for mass propagation [1,2] and phytosanitary problems. In vitro culture (micropropagation) can overcome the limitations of traditional cuttings, presenting an alternative for faster growth [31]. However, one of the first steps to overcome in this process is the establishment of new explants in vitro. Thereby, 'Delite' rabbiteye blueberry was cultured in vitro with four cytokinins: zeatin (ZEA), 6-(γ-γ-dimethylallylamino)-purine (2iP), 6-benzylaminopurine (BAP) and kinetin (KIN) at eight concentrations (0, 2.5, 5, 10, 20, 30, 40 and 50 μM) aimed to establish new explants. Additionally, nine combinations of nitrogen salts were tested, using Woody Plant Medium (WPM) and a modified WPM as the basic medium.

Based on the several combinations of cytokinins and nitrogen salts evaluated in this study, the authors recommended, as an efficient strategy for the in vitro establishment of 'Delite' rabbiteye blueberry, the lowest ZEA concentration (2.5 μM), which promoted a high survival rate (89.7%) as well as a good response of explants forming new shoots (81.3%). This concentration yielded 1.3 new shoots per explant, a high shoot length (13.8 mm) and 10.0 leaves per shoot. Concerning salt composition, the authors recommend the original WPM. An increase or decrease in the NH_4NO_3 and $Ca(NO_3)_2$ concentration did not promote better growth than the original medium.

The fourth contribution was "Pecan Propagation: Seed Mass as a Reliable Tool for Seed Selection" by Poletto et al. [32]. High-quality seedlings are essential in establishing healthy and productive orchards, and studies on new technologies for pecan (*Carya illinoinensis* (Wangenh) K. Koch) cultivation in subtropical regions have mostly focused on seedling production [33,34]. However, seed sizing is a single technique suggested as an efficient indicator positively correlated with seed physiological quality and plantlet vigor [35,36]. In this context, this study evaluated the correlation between seed mass, seed emergence and plantlet vigor in the pecan cultivar Importada, towards producing healthy seedlings for orchard establishment under subtropical climatic conditions.

Among the results reported by the authors, a significant positive correlation between seed mass and plantlet height, stem diameter, emergence rate and number of leaves was observed. Thus, the authors suggested that seed mass can be used as a direct method for seed selection towards production of vigorous pecan seedlings. Therefore, controlled studies—taking into consideration the recalcitrant nature of pecan seeds [37] and the influence of genetic stratification of seed parents on these traits—should be carried out. In addition, an increase in seed mass is usually associated with a decrease in the number of seeds that a plant can produce per unit canopy per year [38].

2.2. Vegetable Crops

The first contribution in this section explores the topic "A Susceptible Scion Reduces Rootstock Tolerance to *Ralstonia solanacearum* in Grafted Eggplant" by Huang et al. [39]. The use of tolerant rootstocks for grafting eggplant (*Solanum melongena*) varieties is the most effective approach to control bacterial wilt disease, caused by *Ralstonia solanacearum,* an important disease in this crop. However, although a disease-tolerant rootstock can generally improve the tolerance of a susceptible scion [4], the disease tolerance of grafted plants is generally lower than that of self-rooted plants [40,41], which suggests that a susceptible scion may also influence the tolerance level of the rootstock. Thus, the authors aimed in this study to establish different grafted combinations of eggplant and determine if they affect plant growth and decrease the bacterial wilt tolerance of a rootstock upon inoculation with *R. solanacearum.*

The authors found that the use of a susceptible scion in scion/rootstock eggplant grafts contributed to a reduction in rootstock tolerance to *R. solanacearum.* Similar results were previously reported for other plant species, including *Capsicum anuum* L. and *Solanum lycopersicum* [42,43], where the use of a susceptible scion on a tolerant rootstock reduced the biological yield of plants to a certain extent when infected with *R. solanacearum.* This phenomenon was documented for eggplant through this study.

The second paper concerned "Aeroponic Cloning of *Capsicum* spp." by Del Valle-Echevarria et al. [3]. Clonally-propagated pepper hybrids would provide an effective way to immortalize favored hybrid genotypes. There are several cloning techniques available, such as in vitro propagation, which presents some disadvantages. In contrast, aeroponic cloning promises to be a great technique that can be used by pepper enthusiasts since it consists of soilless culture that allows for plant growth in a controlled environment, such as a greenhouse or growth chamber that can be readily available [44]. However, peppers have not been evaluated using this technique. Thus, this study evaluated the suitability of an aeroponic cloning protocol by using five *Capsicum* spp. cultivars as well as one closely related wild species.

Interesting findings were reported in this study: All domestic species were successfully regenerated under aeroponic conditions but not *Capsicum eximium*, the wild species. In this context, the success of cloning in all five domesticated *Capsicum* spp. in aeroponic conditions illustrated that it is a viable option for increasing populations of plants with desirable phenotypic traits for both home and boutique growers. Of the species analyzed, *Capsicum annuum* peppers had the fastest node formation and obtained a larger volume of roots after node formation as compared to *C. baccatum, C. frutescens* and *C. pubescens.* This study presented a cost-effective strategy to clonally propagate peppers for personal, industrial and conservation purposes.

2.3. Ornalmental and Medicinal Crops

The first contribution in this section explored "In Vitro Propagation and Acclimatization of Dragon Tree (*Dracaena draco*)" by Galus et al. [45]. *Dracena draco,* known as the dragon tree, is a subtropical ornamental species that presents important interest due to its medicinal properties [46], which is prompting innovative approaches for its efficient use in the food industry and in pharmacological applications. In this study, an efficient in vitro procedure was developed for bud induction, rooting of developing shoots and greenhouse acclimatization of young plantlets of dragon tree.

The MS medium [47] was supplemented with different combinations of kinetin (KIN) or 6-benzylaminopurine (BAP) and naphthaleneacetic acid (NAA) or indole-3-butyric acid (IBA). Media were (S1) 1 mg/L KIN and 1 mg/L NAA; (S2) 3 mg/L KIN and 1 mg/L IAA; (S3) 1 mg/L BAP and 2 mg/L IBA; and (S4) 1 mg/L BAP and 1 mg/L NAA [48], and were used to initiate shoot development on dormant buds recovered from mature *D. draco* trees. Developing shoots were sub-cultured on MS medium containing different concentrations of IBA: (R1) 0 mg/L IBA; (R2) 0.5 mg/L IBA; (R3) 1 mg/L IBA; and (R4) 2 mg/L IBA.

In this study the authors reported that the best shoot induction and rooting media were S1 and S2, and R3 and R4, respectively. Dormant buds from one-year-old *D. draco* plants submitted to this in vitro procedure allowed successful recovery of up to eight individuals per explant used. In vitro grown plants were successfully acclimated in the greenhouse. The potential of this in vitro procedure for multiplication of this endangered tree is discussed in this report.

The second paper in this section is "Improved Propagation and Growing Techniques for Oleander Nursery Production" by Sabatino et al. [49]. Implementing propagation methods to enhance transplant success, establishment and post-plant maintenance is a major objective for plant nurseries involved in the production of shrubs to be used for gardens and natural landscapes in regions with a Mediterranean climate. In this regard, the production of oleander (*Nerium oleander*) rooted cuttings with a well-developed root system is fundamental for successful transplanting and establishment in the field [50]. In the first trial, the authors examined rooting of stem cuttings in relation to number of nodes and indole-3-butyric acid (IBA) treatment in five *N. oleander* clones (clones 1 to 5) grown in Sicily. In a second trial, they tested the effect of different forcing dates and shading on oleander plants for gardens and natural landscapes using just one clone (clone 1).

Interesting results were recorded in both trials: three- and four-node cuttings, ranging in length from 10 to 14 cm, were significantly superior to two-node cuttings (8–10 cm long) in terms of rooting percentage and number of roots per cutting. Concerning IBA application, this treatment improved rooting percentage and root number as compared to an untreated control. However, irrespective of IBA application, rooting percentages ranged from 52% to 94% in clones 4 and 1, respectively. With respect to the shading periods, shaded plants forced in October were significantly higher than those forced in November and in December. Beginning of flowering was delayed in unforced plants. Plants forced in October flowered significantly sooner (first decade of March) than unforced ones (first decade of May) and reached complete flowering almost two months earlier (last week of March). Shading might also save a considerable amount of water during the driest months, especially in Mediterranean areas where water is a scarce resource. Finally, by forcing oleander to flower earlier, growers can supply blooming plants to the market when they would naturally still be vegetative. Overall, this would provide nurseries involved in ornamental plant production with a great marketing tool.

The last contribution examined is "LEDs Combined with CHO Sources and CCC Priming PLB Regeneration of *Phalaenopsis*" by Mehraj et al. [51]. Regeneration of protocorm-like bodies (PLB) is the best and most efficient technique for orchid micropropagation [52], because it has a rapid proliferation capacity for producing a large number of PLBs within a short period [53]. However, environmental factors such as culture media and light source can affect PLB regeneration and plantlet development. Thus, this study determined the best carbohydrate (CHO) source and LED light combination for successful PLB regeneration of *Phalaenopsis* "Fmk02010". In addition, the authors also assessed the impact of chlorocholine chloride (CCC) priming in in vitro PLB propagation of *Phalaenopsis*. The authors applied 15 LEDs combined with three CHO sources and five CCC concentrations to assess the PLB organogenesis.

Among the results, the authors concluded that sucrose and trehalose can be used as excellent CHO sources in the culture media for PLB regeneration of *Phalaenopsis*. Red/white LED + sucrose was the best combination to produce the maximum number of PLBs. However, the combination of blue/white LED + trehalose also produced a large number with healthier PLBs; it also had a tendency to produce a greater number of shoots that would need immediate subculture for future preservation. Red/blue/white LED + trehalose generated a satisfactory number of PLBs with a higher fresh weight and did not generate any shoots. An excessive concentration of CCC ($10~\mathrm{mgL^{-1}}$) caused an enormous reduction in the number of PLBs, the percentage of PLB formation and fresh weight; the addition of low concentrations of CCC in the plant culture medium was also unnecessary.

3. Conclusions

The papers of the Special Issue on "Innovation in Propagation of Fruit, Vegetable and Ornamental Plants" represent innovative research results in this area, with strong applications regarding nursery production and breeding of these species. We hope that this Special Issue will stimulate further research in this area worldwide.

Author Contributions: The contribution to the programming and executing of this Special Issue must be equally divided by S.R.R. and R.C.C. All authors have read and agreed to the published version of the manuscript.

Funding: This research received no external funding.

Acknowledgments: We thank all authors of the Special Issue.

Conflicts of Interest: The authors declare no conflicts of interest.

References

1. Marino, S.R.; Williamson, J.G.; Olmstead, J.W.; Harmon, P.F. Vegetative growth of three southern highbush blueberry cultivars obtained from micropropagation and softwood cuttings in two Florida locations. *HortScience* **2014**, *49*, 556–561. [CrossRef]
2. Meiners, J.; Schwab, M.; Szankowski, I. Efficient in vitro regeneration systems for *Vaccinium* species. *Plant Cell Tissue Organ Cult.* **2007**, *89*, 169–176. [CrossRef]
3. Del Valle-Echevarria, A.R.; Kantar, M.B.; Branca, J.; Moore, S.; Frederiksen, M.K.; Hagen, L.; Hussain, T.; Baumler, D.J. Aeroponic Cloning of *Capsicum* spp. *Horticulturae* **2019**, *5*, 30. [CrossRef]
4. King, S.R.; Davis, A.R.; Liu, W.; Levi, A. Grafting for disease resistance. *HortScience* **2008**, *43*, 1673–1676. [CrossRef]
5. Miller, L.R.; Murashige, T. Tissue culture propagation of tropical foliage plants. *In Vitro Cell. Dev. Biol. Anim.* **1976**, *12*, 797–813. [CrossRef]
6. Vinterhalter, D. In vitro Propagation of Green-Foliaged *Dracaena-Fragrans* Ker. *Plant Cell Tissue Org.* **1989**, *17*, 13–19.
7. Vinterhalter, D.; Vinterhalter, B. Micropropagation of Dracaena Species. In *Biotechnology in Agriculture and Forestry*; Springer: Berlin/Heidelberg, Germany, 1997; Volume 40, pp. 131–146.
8. Tian, L.; Tan, H.Y.; Zhang, L. Stem segment culture and tube propagation of *Dracaena saneriana* cv. virscens. *Acta Hortic. Sin.* **1999**, *26*, 133–134.
9. Blanco, M.; Valverde, R.; Gomez, L. Micropropagation of *Dracaena deremensis*. *Agron. Costarric.* **2004**, *28*, 7–15.
10. Koyama, R.; Aparecido Ribeiro Júnior, W.; Mariani Zeffa, D.; Tadeu Faria, R.; Mitsuharu Saito, H.; Simões Azeredo Gonçalves, L.; Ruffo Roberto, S. Association of indolebutyric acid with *Azospirillum brasilense* in the rooting of herbaceous blueberry cuttings. *Horticulturae* **2019**, *5*, 68. [CrossRef]
11. Mariosa, T.N.; Melloni, E.G.P.; Melloni, R.; Ferreira, G.M.R.; Souza, S.M.P.; Silva, L.F.O. *Rhizobacteria* and development of seedlings from semi-hardwood cuttings of olive (*Olea europaea* L.). *Revista de Ciências Agrárias* **2017**, *60*, 302–306. [CrossRef]
12. Khademian, R.; Asghari, B.; Sedaghati, B.; Yaghoubian, Y. Plant beneficial rhizospheric microorganisms (PBRMs) mitigate deleterious effects of salinity in sesame (*Sesamum indicum* L.): Physio-biochemical properties, fatty acids composition and secondary metabolites content. *Ind. Crops. Prod.* **2019**, *136*, 129–139. [CrossRef]
13. Cassán, F.; Diaz-Zorita, M. *Azospirillum* sp. in current agriculture: From the laboratory to the field. *Soil Biol. Biochem.* **2016**, *103*, 117–130. [CrossRef]
14. Molina, R.; Rivera, D.; Mora, V.; López, G.; Rosas, S.; Spaepen, S.; Vanderleyden, J.; Cassán, F. Regulation of IAA biosynthesis in *Azospirillum brasilense* under environmental stress conditions. *Curr. Microbiol.* **2018**, *75*, 1408–1418. [CrossRef] [PubMed]
15. Rivera, D.; Mora, V.; Lopez, G.; Rosas, S.; Spaepen, S.; Vanderleyden, J.; Cassan, F. New insights into indole-3-acetic acid metabolism in *Azospirillum brasilense*. *J. Appl. Microbiol. Biochem.* **2018**, *125*, 1774–1785. [CrossRef]
16. Fukami, J.; Ollero, F.J.; de la Osa, C.; Valderrama-Fernández, R.; Nogueira, M.A.; Megías, M.; Hungria, M. Antioxidant activity and induction of mechanisms of resistance to stresses related to the inoculation with *Azospirillum brasilense*. *Arch. Microbiol.* **2018**, *200*, 1191–1203. [CrossRef]
17. Gradziel, T.; Lampinen, B.; Preece, J.E. Propagation from basal epicormic meristems remediates an aging-related disorder in almond clones. *Horticulturae* **2019**, *5*, 28. [CrossRef]
18. Kester, D.E. The clone in Horticulture. *HortScience* **1983**, *18*, 831–837.
19. Skowcroft, W.R. Somaclonal Variation, the Myth of Clonal Uniformity. In *Genetic Flux in Plants*; Hohn, B., Dennis, E.S., Eds.; Springer: New York, NY, USA, 1985; pp. 217–245.
20. Skirvin, R.M.; McPheeters, K.D.; Norton, M. Sources and frequency of somaclonal variation. *HortScience* **1994**, *29*, 1232–1237. [CrossRef]
21. D'Aquila, P.; Rose, G.; Bellizzi, D.; Passarino, G. Epigenetics and aging. *Maturitas* **2013**, *74*, 130–136. [CrossRef]

22. Fraga, M.F.; Rodriguez, R.; Canal, M.J. Genomic DNA methylation-demethylation during aging and reinvigoration of *Pinus radiata*. *Tree Physiol.* **2002**, *22*, 813–816. [CrossRef]
23. Kester Dale, E.; Asay, R.N. Variability in noninfectious bud-failure of 'Nonpareil' almond. Location and environment. *J. Am. Soc. Hortic. Sci.* **1978**, *103*, 377–382.
24. Fenton, C.A.L.; Kuniyuki, A.H.; Kester, D.E. Search for a viroid etiology for noninfectious bud failure in almond. *HortScience* **1988**, *23*, 1050–1053.
25. Kester, D.E.; Gradziel, T.M. Genetic Disorders. In *Almond Production Manual*; Micke, W.C., Ed.; University of California: Oakland, CA, USA, 1996; pp. 76–87.
26. Kester, D.E. Noninfectious Bud-Failure in Almond. In *Virus Diseases and Disorders of Stone Fruits in North America*; Fulton, R.W., Ed.; Agricultural Research Service, U.S. Department of Agriculture: Washington, DC, USA, 1976; pp. 278–283.
27. Kester, D.E. Noninfectious bud-failure, a nontransmissable inherited disorder in almond. I. Pattern of phenotypic inheritance. *Proc. Am. Soc. Hortic. Sci.* **1968**, *92*, 7–15.
28. Kester, D.E. Noninfectious bud-failure, a nontransmissable inherited disorder in almond II. Progeny tests for bud-failure. *Proc. Am. Soc. Hortic. Sci.* **1968**, *92*, 16–28.
29. Schuchovski, C.S.; Biasi, L.A. In Vitro establishment of 'Delite' rabbiteye blueberry microshoots. *Horticulturae* **2019**, *5*, 24. [CrossRef]
30. Debnath, S.C. A scale-up system for lowbush blueberry micropropagation using a bioreactor. *HortScience* **2009**, *44*, 1962–1966. [CrossRef]
31. Debnath, S.C. Temporary immersion and stationary bioreactors for mass propagation of true-to-type highbush, half-high, and hybrid blueberries (*Vaccinium* spp.). *J. Hortic. Sci. Biotechnol.* **2017**, *92*, 72–80. [CrossRef]
32. Poletto, T.; Stefenon, V.M.; Poletto, I.; Muniz, M.F.B. Pecan propagation: Seed mass as a reliable tool for seed selection. *Horticulturae* **2018**, *4*, 26. [CrossRef]
33. Cargnelutti Filho, A.; Poletto, T.; Muniz, M.F.B.; Baggiotto, C.; Poletto, I.; Fronza, D. Sampling design for height and diameter evaluation of pecan seedlings. *Ciênc. Rural* **2014**, *44*, 2151–2156. [CrossRef]
34. Poletto, I.; Muniz, M.F.B.; Poletto, T.; Stefenon, V.M.; Baggiotto, C.; Ceconi, D.E. Germination and development of pecan cultivar seedlings by seed stratification. *Pesqui. Agropecu. Bras.* **2015**, *50*, 1232–1235. [CrossRef]
35. Pereira, W.A.; Pereira, S.M.A.; Dias, D.C.F.S. Influence of seed size and water restriction on germination of soybean seeds and on early development of seedlings. *J. Seed Sci.* **2013**, *35*, 316–322. [CrossRef]
36. Bispo, J.S.; Costa, D.C.C.; Gomes, S.E.V.; Oliveira, G.M.; Matias, J.R.; Ribeiro, R.C.; Dantas, B.F. Size and vigor of *Anadenanthera colubrina* (Vell.) Brenan seeds harvested in Caatinga areas. *J. Seed Sci.* **2017**, *39*, 363–373. [CrossRef]
37. Dalkiliç, Z. Effects of drying on germination rate of pecan seeds. *J. Food Agric. Environ.* **2013**, *11*, 879–882.
38. Henery, M.L.; Westoby, M. Seed mass and seed nutrient content as predictors of seed output variation between species. *Oikos* **2001**, *92*, 479–490. [CrossRef]
39. Huang, C.; Wang, Y.; Yang, Y.; Zhong, C.; Notaguchi, M.; Yu, W. A Susceptible scion reduces rootstock tolerance to *Ralstonia solanacearum* in grafted eggplant. *Horticulturae* **2019**, *5*, 78. [CrossRef]
40. Nakaho, K.; Inoue, H.; Takayama, T.; Miyagawa, H. Distribution and multiplication of *Ralstonia solanacearum* in tomato plants with resistance derived from different origins. *J. Gen. Plant Pathol.* **2004**, *70*, 115–119. [CrossRef]
41. Seemüller, E.; Harries, H. *Plant Resistance. Phytoplasmas: Genomes, Plant Hosts and Vectors*; CAB International: Oxfordshire, UK, 2010; pp. 147–169.
42. Liu, Y.; Jiang, F.; Zhang, N.; Wang, H.; Ai, X. Relationship between osmoregulation and bacterial wilt resistance of grafted pepper. *Acta Hortic. Sin.* **2011**, *38*, 903–910.
43. McAvoy, T.; Freeman, J.H.; Rideout, S.L.; Olson, S.M.; Paret, M.L. Evaluation of grafting using hybrid rootstocks for management of bacterial wilt in field tomato production. *HortScience* **2012**, *47*, 621–625. [CrossRef]
44. Ritter, E.; Angulo, B.; Riga, P.; Herran, C.; Relloso, J.; San Jose, M. Comparison of hydroponic and aeroponic cultivation systems for the production of potato minitubers. *Potato Res.* **2001**, *44*, 127–135. [CrossRef]
45. Galus, A.; Chenari Bouket, A.; Belbahri, L. In Vitro propagation and acclimatization of Dragon Tree (*Dracaena draco*). *Horticulturae* **2019**, *5*, 64. [CrossRef]
46. Jura-Morawiec, J.; Tulik, M. Dragon's blood secretion and its ecological significance. *Chemoecology* **2016**, *26*, 101–105. [CrossRef] [PubMed]

47. Murashige, T.; Skoog, F. A Revised Medium for Rapid Growth and Bio Assays with Tobacco Tissue Cultures. *Physiol. Plant.* **1962**, *15*, 473–497. [CrossRef]
48. Liu, J.; Deng, M.; Henny, R.J.; Chen, J.; Xie, J. Regeneration of *Dracaena surculosa* through indirect shoot organogenesis. *HortScience* **2010**, *45*, 1250–1254. [CrossRef]
49. Sabatino, L.; D'Anna, F.; Iapichino, G. Improved propagation and growing techniques for oleander nursery production. *Horticulturae* **2019**, *5*, 55. [CrossRef]
50. Pilon, P. *Perennial Solutions: A Grower's Guide to Perennial Production*, 1st ed.; Ball Publishing: Batavia, IL, USA, 2005; p. 546.
51. Mehraj, H.; Alam, M.M.; Habiba, S.U.; Mehbub, H. LEDs Combined with CHO sources and CCC priming PLB regeneration of *Phalaenopsis*. *Horticulturae* **2019**, *5*, 34. [CrossRef]
52. Arditti, J.; Ernst, R. *Micropropagation of Orchids*; Wiley: New York, NY, USA, 1993; pp. 1–682.
53. Sheelavanthmath, S.S.; Murthy, H.N.; Hema, B.P.; Hahn, E.J.; Paek, K.Y. High frequency of protocorm like bodies (PLBs) induction and plant regeneration from protocorm and leaf sections of *Aerides crispum*. *Sci. Hortic.* **2005**, *106*, 395–401. [CrossRef]

 *horticulturae*

Article

A Susceptible Scion Reduces Rootstock Tolerance to *Ralstonia solanacearum* in Grafted Eggplant

Chaokun Huang [1,2,†], **Yuexia Wang** [1,†], **Yanjuan Yang** [1], **Chuan Zhong** [1], **Michitaka Notaguchi** [2] **and Wenjin Yu** [1,*]

[1] Horticultural Science, Agricultural College of Guangxi University, Nanning 530004, China; hck5632@icloud.com (C.H.); wangyueya33@163.com (Y.W.); yjyang85@126.com (Y.Y.); zhongchuan11@hotmail.com (C.Z.)

[2] Bioscience and Biotechnology Center, Nagoya University, Nagoya 4648601, Japan; notaguchi.michitaka@b.mbox.nagoya-u.ac.jp

[*] Correspondence: yuwjin@gxu.edu.cn

[†] These authors contributed equally to this work.

Received: 29 July 2019; Accepted: 12 December 2019; Published: 12 December 2019

Abstract: The bacterial wilt pathogen (*Ralstonia solanacearum*) is a highly pathogenic soil-borne bacterium that invades the vascular system of a host plant leading to plant wilting and death. In agricultural systems, tolerant rootstocks are usually used to enhance disease resistance and tolerance in crop plants to soil-borne pathogens. Here, two distinct eggplant cultivars with different tolerances to *R. solanacearum* infection, the disease-tolerant cultivar 'S21' and the disease-susceptible cultivar 'Rf', were used to investigate if scion tolerance level can affect tolerance of rootstock upon an infection of the same pathogen. Three scion/rootstock grafted combinations were considered: Rf/S21, S21/S21, and Rf/Rf. Plants that resulted from the combination Rf/S21, composed of the susceptible scion grafts, showed weak tolerance to *R. solanacearum* infection, and exhibited the poorest growth compared to the tolerant scion grafts (S21/S21). As expected, the combination Rf/Rf showed the lowest level of disease tolerance. Furthermore, a high level of exopolysaccharides (EPSs) and cell wall degrading enzymes (CWDEs) were detected in susceptible scion grafts. These factors are involved in plant growth inhibition due to blocking transport between scion and rootstock and damage of vascular tissues in the plant. A high level of reactive oxygen species (ROS) and active oxygen scavenging enzymes were also detected in susceptible scion grafts. Excess accumulation of these substances harms the dynamic balance in plant vascular systems. These results indicated that the use of a susceptible scion in scion/rootstock eggplant grafts contributed to a reduction in rootstock tolerance to *Ralstonia solanacearum*.

Keywords: bacterial wilt; *Solanum melongena*; susceptible; tolerance; exopolysaccharides; cell wall degrading enzymes

---

1. Introduction

Bacterial wilt is a soil-borne disease of eggplant (*Solanum melongena*) caused by *Ralstonia solanacearum E.F. Smith*, which usually occurs in tropical, subtropical, and temperate regions (Smith, 1896). The bacterial wilt pathogen enters the host plant through the roots and invades the vascular system [1,2]. *R. solanacearum* infects xylem parenchyma and stimulates cells to form invasive compounds in the vicinity of the vascular tissues, which are then released into the vessels and sieve tubes. The pathogen rapidly proliferates and spreads in vascular tissues, causing obstruction of the xylem and phloem [3]. *R. solanacearum* is also similar to most plant pathogenic bacteria that can secrete various toxic proteins that induce a hypersensitive response in the host plant and activate a complex plant defense network [4].

Studies have shown that *R. solanacearum* can produce a large number of exopolysaccharides (EPSs) in the vascular tissues, which block xylem and phloem [5]. EPSs hinder water and nutrient transport in the plant's vascular system, resulting in wilting symptoms, and therefore EPSs are an important pathogenicity factor of bacterial wilt disease [6,7]. *R. solanacearum* secretes cell wall degrading enzymes (CWDEs) during host infection, including cellulase (Cx), pectin methylgalacturonase (PMG), and pectin methyl transelimination enzyme (PMTE), which have been studied for their potential in promoting pathogenesis and causing disease symptoms [8–10]. In general, CWDEs produced by plant pathogens are considered to be important pathogenicity factors. Significant work has been published on the role of CWDEs during plant pathogen infection, and studies show that the accumulation of CWDSs will accelerate the decline of plant cells [11,12].

Once host plants are infected with *R. solanacearum*, pathogens cannot be directly inhibited or eliminated, therefore plants induce and excrete defense factors including phytoalexin and reactive oxygen species (ROS) that enhance defense against pathogens [13]. Superoxide anion (O_2^-) and hydrogen peroxide (H_2O_2) are the major forms of ROS. These molecules are highly reactive and toxic and can lead to the oxidative destruction of cells. The rapid accumulation of plant ROS at the pathogen attack site, a phenomenon called oxidative burst, is directly toxic to pathogens and can lead to a hypersensitive response that results in a zone of host cell death, which prevents the further spread of biotrophic pathogens [14,15]. However, a greater degree of *R. solanacearum* stress on the plant tissues results in the excess accumulation of ROS, which increases the burden of the active oxygen-scavenging system in plants [16]. Active oxygen-scavenging systems in plants mainly include superoxide dismutase (SOD), catalase (CAT), peroxidase (POD), and ascorbate peroxidase (APX) [17,18]. The main role of SOD is to remove the O_2^- present in the cytoplasm, chloroplasts, and peroxisomes to prevent oxygen radicals from damaging the cell membrane system [19]. CAT and POD facilitate H_2O_2 enzymatic degradation to prevent an over-accumulation of more toxic OH^- [20]. APX is also a H_2O_2 scavenger that uses the APX-AsA cycle to reduce damage from H_2O_2 in plant cells [21,22]. These enzymes, to a certain extent, have become the protective system for plants facing environmental stress, maintaining the balance of ROS metabolism within the cells and protecting the membrane structure. However, plant cells will also be damaged if the excess accumulation of ROS exceeds the ability of the active oxygen scavenging enzyme system [23].

At present, the use of tolerant rootstocks for grafting eggplant varieties is the most effective approach to control bacterial wilt disease [24]. There are complex interactions between the rootstock and scion during the period of initiation and the completion of healing in grafts [25]. Investigations on the interaction of rootstock and scion have mostly been focused on physiological and biochemical characteristics regarding the mechanisms involved in graft compatibility, nutrient and water uptake, assimilation and translocation of solutes, and the influence of the rootstock on the main physiological processes of the scion [26]. Studies have shown that rootstocks not only affect the nutritional status, water metabolism, and hormonal signaling of the scion, but they can also increase the resistance and tolerance of the scion to disease and environment stresses, resulting in a higher yield [27,28]. Although a disease-tolerant rootstock can generally improve the tolerance of a susceptible scion [24], the disease tolerance of grafted plants is generally lower than that of self-rooted plants [29,30], which suggests that a susceptible scion may also influence the tolerance level of the rootstock. However, to our knowledge, there are no studies to determine if the disease tolerance level of a scion directly affects rootstock tolerance to a specific disease.

In this study, different grafted combinations of eggplant were established to determine if they affect plant growth and abate the bacterial wilt tolerance of a rootstock upon inoculation with *R. solanacearum*.

2. Materials and Methods

2.1. Plant Material and Ralstonia Solanacearum

Eggplant genotypes 'Qiezhen No. 21' (S21, high tolerance to bacterial wilt), bred by Guangxi University, Nanning, China, and "Zichang eggplant No. 1" (Rf, high susceptibility to bacterial wilt), a commercial variety, were used to establish the experiment. The grafted scion/rootstock combinations Rf/S21, S21/S21, and Rf/Rf were constructed as the experimental plants.

R. solanacearum (bacterial wilt pathogen) was provided by the Plant Pathology Laboratory of the Agricultural College at Guangxi University. The bacterium was mixed with a 30% glycerol solution and stored at −80 °C. Before inoculation, metal rods were dipped into the stored *R. solanacearum* and were used to draw several lines on sterile nutrient agar (NA) medium. The stored pathogens were transferred to the medium and were activated and incubated in a 28 °C incubator. After 2 d of incubation, media with no contamination and with good bacterial growth (a single white colony, strongly pathogenic in subsequent experiments) (Figure 1) were selected as inocula.

Figure 1. (**a**) Grafting seedlings with C-type annular tubes. (**b**) Pathogenic *R.solanacearum* in NA medium. (**c**) Disease classification standard for the plant.

2.2. Grafting Method, Grafted Plant Cultivation and R. solanacearum Inoculum

Eggplant seeds of each graft combination and a non-grafted Rf control were soaked in warm water (50 °C) and sown in 50-hole trays with commercial plant cultivation substrate to promote germination. When seedlings exhibited two true leaves, they were grafted with C-type annular tubes (Figure 1) using the close joining method [31]. The grafted seedlings were then kept in a growth chamber under controlled conditions (24–28 °C, 90–95% relative humidity, darkness) for 12 h. Further, to promote graft healing, the temperature of the growth chamber was changed to 20–22 °C for 3 d, then seedlings were moved to the greenhouse (26–28 °C). After 15 d of growth under greenhouse conditions, the grafts were transplanted into larger pots containing substrate. Physiological parameters (see in detail below) were measured during this period. When the grafted plants exhibited 4–5 true leaves, *R. solanacearum* inoculum, diluted in water to a concentration of $OD_{600} = 0.5$, was applied using the root damage infused inoculum method [32]. After inoculation, seedlings were incubated in a growth chamber for

28 d under a 14 h photoperiod with 30 °C day/25 °C night and 70% relative humidity. Non-inoculated plants of each graft combination were included and otherwise handled the same way. Prepare enough experimental plants according to the experimental arrangement (Table S1). Containers were watered and fertilized twice a week from seed sowing to the end of the experiment, approximately 2 to 3 months.

2.3. Disease Classification and Disease-Tolerance Evaluation

To identify disease symptoms, plants of the graft combinations and the non-grafted control were analyzed at 28 d post-inoculation (DPI) with *R. solanacearum*. Twenty plants were evaluated per grafted combination and control and repeated three times (80 plants per repetition, for a total of 240 plants, were considered; Table S1). The level of disease tolerance was evaluated at 28 DPI following the procedure described by Liu et al. [33]. Five different levels of plant tolerance to *R. solanacearum* infection were defined: Grade 0: no obvious symptoms; Grade 1: one leaf wilting; Grade 2: 2–3 leaves wilting; Grade 3: four leaves wilting; Grade 4: whole seedling wilting or death (Figure 1). The parameters, incidence rate (IR) and disease index (DI) were determined according to the mathematical expressions below. Incidence rate (IR) = number of infected plants/number of inoculated plants. Disease index (DI) = $\sum$ (disease grade × number of infected plants)/(the highest disease grade × number of inoculated plants) × 100. Five levels of plant tolerance to bacterial wilt were considered (based on the DI parameter): I: immune (DI = 0); HT: highly tolerant (0 < DI ≤ 15); T: tolerant (15 < DI ≤ 30); MT: moderately tolerant (30 < DI ≤ 45); S: susceptible (45 < DI ≤ 60); HS: highly susceptible (DI > 60) [9]. The experiment was a completely randomized design with three repetitions, analysis of variance (ANOVA) was used to test for significance, and significant differences ($P < 0.05$) between treatments were determined using Tukey's test.

2.4. Measurement of Plant Growth Situation

Ten plants from each graft combination were randomly selected after disease-tolerance evaluation (30 plants per experimental repetition, a total of 90 plants upon 3 repetitions; Table S1) at 28 DPI. The biomass that reflects plant growth was measured, including plant height (g), stem diameter (cm), root length (cm), and surface area (cm^2), and dry and fresh weight (g) of the plants. Plant height was considered as length between the base of the stem and the end of the apical meristem. Plant height was measured by a straightedge ruler. The stem diameter of the scion was measured approximately 0.5 cm above the graft interface and stem diameter of the rootstock was measured approximately 0.5 cm below the graft interface, both using a Vernier caliper. After washing roots to remove media, root length and surface area were measured using a Founder Z2400 scanner and analyzed with WinRHIZO 2009c software. Fresh weights of plant stems and roots were measured with an electronic Mettler Toledo balance. For dry weight determination, seedlings were placed in a ETTAS ON-300S drying oven, at 105 °C for 15 min for enzyme inactivation, dried at 45 °C until a constant weight, and then measured with the electronic balance. To calculate normalized values in this experiment, the biomass measurement values at 28 DPI minus their average value at 0 DPI (average from 10 plants per grafted combination), were calculated.

2.5. Determination of EPS Content and Cell Wall Degrading Enzyme (CWDE) Activity of Grafted Plants

Samples were collected from grafted plants at 0, 7, 14, and 28 DPI with *R. solanacearum*. Ten plants were selected from each grafted combination at each experimental timepoint (30 plants per grafted combination at each time point, for a total of 360 plants upon 3 repetitions; Table S1). Samples taken from stems (0.2 g) and roots were individually ground well with a mortar and pestle and diluted with 10 mL distilled water. After homogenization, samples were boiled for 30 min, centrifuged at 12,000× g for 10 min at 25 °C, and supernatants collected into a 25 mL bottle. These supernatants were then used to measure EPS content following the phenol-sulfuric acid method described by Cao et al. [34].

Briefly, 0.5 mL of supernatant solution was mixed with 1.5 mL of distilled water and after mixing well, 1.0 mL of phenol solution (90 g·L^{-1}) and 5.0 mL concentrated sulfuric acid were added. After 30 min of incubation at room temperature, absorbance was measured at a 485 nm wavelength.

To proceed with CWDE activity measurements, extraction buffer was prepared by dissolving 11.78 g NaCl in a 250 mL sodium acetate–acetic acid buffer solution (0.05 M, pH 5.5). Afterwards, samples (0.2 g) taken from stems and roots were individually thoroughly ground with a mortar and pestle and diluted with 1.6 mL extraction buffer in an ice bath. Then, centrifugation was performed at 12,000× *g* and 4 °C for 20 min, and supernatants (extracted solution) were collected for measuring CWDE activity [35].

To measure the cellulase (Cx) activity, the protocol described by Li et al. [35] was followed. Briefly, 0.1 mL of extracted solution was added to 0.2 mL carboxymethylcellulose sodium (CMC) solution (0.6% *v/v* concentration), with a 30 min incubation in a 50 °C water bath. Then, 1.5 mL 3,5-dinitrosalicylic acid (DNS) solution (6.3 g/L) was immediately added to stop the reaction. The absorbance was measured at a 485 nm wavelength.

To determine the pectin methyl-galacturonase (PMG) activity, the protocol previously described by Li et al. [35] was followed. Briefly, 0.1 mL of extracted solution was added to a solution consisting of 0.1 mL pectin solution (0.25% *v/v* concentration) and 0.3 mL acetate buffer (0.05 M, pH 5.5). Afterwards, 1.5 mL DNS solution (6.3 g/L) was added to stop the reaction. The absorbance was measured at a 540 nm wavelength.

The method used to determine the pectin methyl trans elimination enzyme (PMTE) activity followed the protocol described by Li et al. [35], consisting of addition of 0.1 mL of extracted solution to the previous prepared mixture of 0.1 mL pectin solution (0.25% concentration, containing 0.6 mmol·L^{-1} of CaCl$_2$·2H$_2$O) and 0.3 mL glycine–sodium hydroxide buffer (0.05 M, pH 9.0). The absorbance was measured at a 232 nm wavelength.

For calculating relative enzyme activity (U) in each of the assays above, the absorbance changes of the extracted solution were recorded over 1 min.

2.6. Determination of ROS and Active Oxygen Scavenging Enzymes Activity in Grafted Plants

The level of ROS and the activity of active oxygen-scavenging enzymes were measured following protocols already described in several reports [36–39]. For those analyses, enzyme extraction buffer (2.5 g polyvinyl pyrrolidone (PVP) in a 100 mL 0.05 M, pH 7.8 phosphate buffer) was prepared prior to plant material processing. Afterwards, samples taken from stems (0.2 g) and roots were individually thoroughly ground with a mortar and pestle and diluted with 2.0 mL extraction buffer in an ice bath. Afterwards, solutions were centrifuged at 12,000× *g* and 4 °C for 30 min and supernatants (extracted solution) were collected for the next experiment.

To determine the content of the superoxide anion (O$_2^-$), the protocol described by Dhindsa et al. [37] was followed. Briefly, 1.0 mL of extracted solution was mixed well with 1.0 mL phosphate buffer (50 mM, pH 7.8) and 1.0 mL hydroxylamine hydrochloride (1 mM) solution. The mixture was incubated at 25 °C for 1 h. Further, 1.0 mL P-aminophenylsulfonic acid (1.0 mmol·L^{-1}) and 1.0 mL α-naphthylamine (7 mmol·L^{-1}) were added. The absorbance was measured at a 530 nm wavelength.

To determine the content of hydrogen peroxide (H$_2$O$_2$), the method described by Dhindsa et al. [37] was followed. Briefly, 1.0 mL of extracted solution was mixed well with 0.4 mL color-developing solution (containing 100 mL of 0.1 M phosphate buffer, 25 μL of N, N-dimethylaniline, 100 mg of 4-aminoantipyryline), and 0.1 mL horseradish peroxidase (250 μg·mL^{-1}). The absorbance was measured at a 550 nm wavelength.

For superoxide dismutase (SOD) activity determination, the method previously reported by Kochba et al. [38] was followed. Briefly, 0.1 mL of previously extracted solution was mixed with 2.6 mL reaction mixture solution (102 mL of 0.05 M, pH 7.0 phosphate buffer; 18 mL of 130 mM DL-methionine (MET); 18 mL of 750 μM nitro blue tetrazolium (NBT); 18 mL of 100 μM EDTA-Na$_2$), and 0.3 mL riboflavin (20 µM). After a 15 min reaction under lights (4000 lx), the reaction was blocked in the dark for 5 min. The absorbance was measured at a 560 nm wavelength.

To determine catalase (CAT) activity, the method described by Kochba et al. [38] was followed, which consisted of mixing 0.1 mL of extracted solution within a 2.9 mL phosphate buffer (0.05 M, pH 7.8) and 0.1 mL H_2O_2 (2% *v/v* concentration). The absorbance was measured at a 240 nm wavelength.

For peroxidase (POD) activity determination, the protocol described by Dalton et al. [39] was followed. Briefly, 0.1 mL of extracted solution was mixed with a 3 mL reaction mixture solution (100 mL of 0.02 M phosphate buffer at pH 6.0, 38 µL of 30% *v/v* H_2O_2, and 0.112 mL of guaiacol). The absorbance was measured at a 470 nm wavelength.

To determine ascorbate peroxidase (APX) activity, the method described by Dalton et al. [39] was followed. Briefly, 0.1 mL of extracted solution was mixed well with 2.6 mL phosphate buffer (0.05 M, pH 7.0, containing 0.1 mM of EDTA-Na$_2$), 150 µL H_2O_2 (20 mM), and 150 µL ascorbic acid (5 mM). The absorbance was measured at a 190 nm wavelength.

For calculating relative enzyme activity (U), the absorbance change value of the extracted solution was recorded over 1 min.

3. Results

3.1. Susceptible Scion Grafts Showed a Weak Tolerance to R. solanacearum

Disease index (DI) and incidence rate (IR) parameters were calculated according to the disease classification standard (Figure 1). The IR of graft S21/S21 was 8.33%, and the DI was 7.23, showing a high tolerance level. The IR of graft Rf/S21 was 26.67%, and the DI was 22.64, showing normal tolerance. The IR of graft Rf/Rf was 100%, and the DI was 97.08, showing high susceptibility (Table 1). The non-grafted plant Rf exhibited a similar level of susceptibility as the Rf/Rf graft, suggesting that grafting did not enhance susceptibility of the susceptible scion (Rf). The disease tolerance level of the different graft combinations was S21/S21 > Rf/S21 > Rf/Rf, which demonstrated that the tolerance level of grafted rootstock decreased by grafting with the susceptible scion (Rf). Thus, the tolerance of a tolerant rootstock was attenuated with the susceptible scion graft (Rf/S21).

Table 1. Tolerance of different grafting combinations to *R. solanacearum*.

Grafted Combination (Scion/Rootstock)	No. Inoculated Plants	No. Infected Plants	Incidence Rate (IR) (%)	Disease Index (DI)	Tolerance Level
S21/S21	60	5	8.33a [z]	7.23a	HT [y]
Rf/S21	60	16	26.67b	22.64b	T
Rf/Rf	60	60	100c	97.08c	HS
Rf (non-grafted)	60	60	100c	96.25c	HS

[z] Different lowercase letters indicate a significant difference ($P < 0.05$) according to Tukey's test. [y] HT indicates high tolerance, T indicates tolerance, and HS indicates high susceptibility; the Rf non-grafted plant was the experimental control.

3.2. Susceptible Scion Grafts Have a Worse Growth Situation Post Inoculation with R. solanacearum

Plant growth parameters, measured 28 days post-inoculation (DPI) with *R. solanacearum*, were used to investigate the effects of the susceptible scion on rootstock tolerance upon infection with the pathogen. There was no significant difference in root length and root surface area between grafts Rf/S21 and S21/S21 in the non-inoculated *R. solanacearum* (Data S1). However, pathogen-inoculated graft Rf/S21 exhibited a shorter root length and less surface area compared to S21/S21 (Figure 2). This negative effect on plant growth was also observed for plant height, stem diameter, and dry and fresh weight (Figures 3 and 4). Without inoculation, root length, surface area, stem diameter, and plant weights were similar for Rf/S21 compared to S21/S21. Because inoculation with *R. solanacearum* inhibited growth of Rf/S21 plants, growth of the tolerant rootstock was abated by the susceptible scion.

These findings suggest that the susceptible scion (Rf) had a negative effect on the eggplant rootstock post-inoculation with *R. solanacearum*.

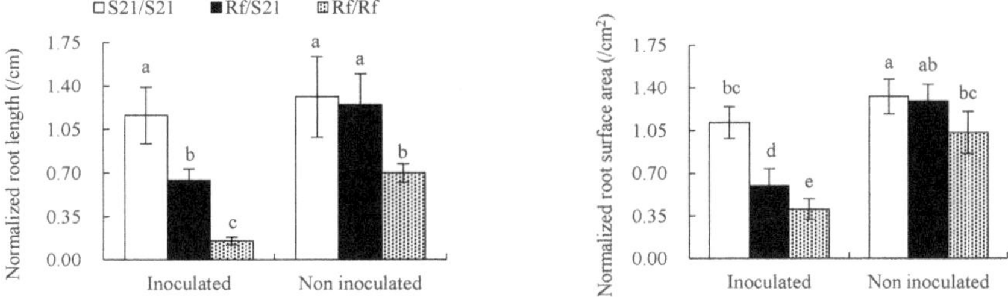

Figure 2. Normalized root length and root surface area in different graft combinations. The error bar indicates the standard error, $n = 30$; the different letters indicate a significant difference ($P < 0.05$) across all combinations according to Tukey's test.

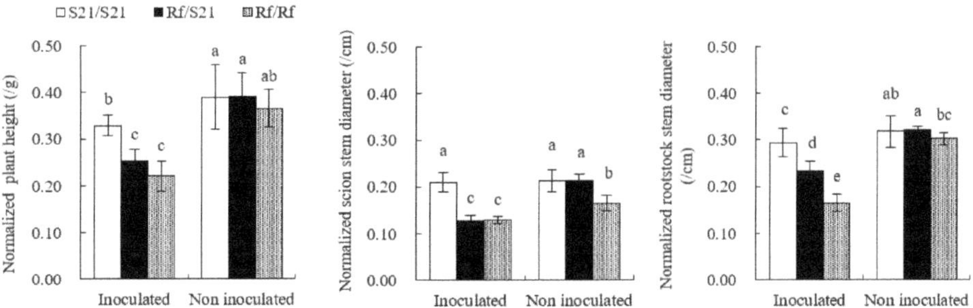

Figure 3. Normalized plant weight, scion stem diameter, and rootstock stem diameter in different graft combinations. Note: The error bar indicates the standard error, $n = 30$; the different letters indicate a significant difference ($P < 0.05$) across all combinations according to Tukey's test.

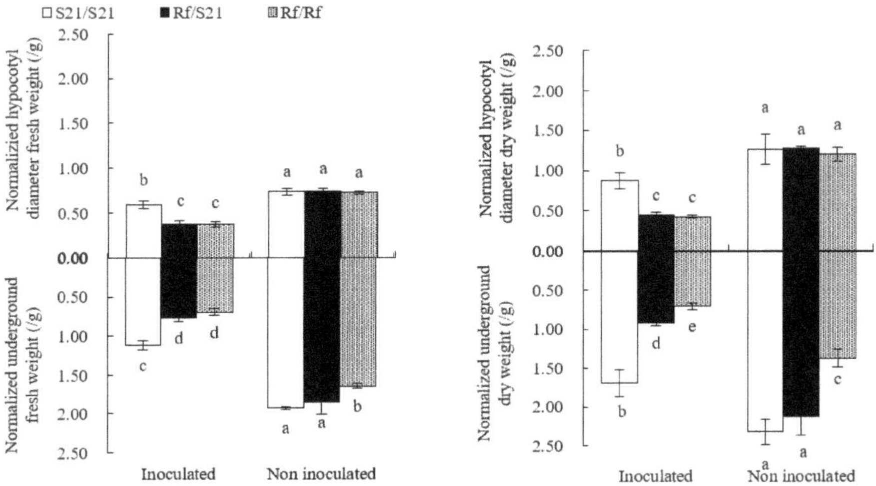

Figure 4. Normalized dry weight and fresh weight of plant stems and roots in different graft combinations. Note: The error bar indicates the standard error, $n = 30$; the different letters indicate a significant difference ($P < 0.05$) across all combinations according to Tukey's test.

3.3. High Levels of EPS and CWDEs Accumulated in Susceptible Scion Grafts

To further investigate the inhibitory effect of the susceptible scion on the tolerant rootstock after infection with bacterial wilt, the exopolysaccharide (EPS) content and cell wall degrading enzyme (CWDE) activities were measured in the rootstock (Data S2). Rf/S21 produced more EPSs than S21/S21 in both stem and roots. The EPS continuously accumulated in Rf/S21 rootstock over time, reaching a peak at 28 DPI with *R. solanacearum* (Figure 5). The EPS content of the S21/S21 rootstock reached a peak value at 14 DPI with *R. solanacearum* and subsequently declined. In Rf/Rf (high disease index, Table 1), the EPS accumulation of the rootstock was much higher than that in Rf/S21 and S21/S21. Similarly, the CWDE (Cx, PMG, and PMTE) activities of Rf/S21 rootstocks were higher than those in the S21/S21 rootstock, both in stem and roots. These enzyme activities also reached their peak value in the S21/S21 rootstock at 14 DPI with *R. solanacearum* and subsequently declined (Figures 6–8). Overall, the EPSs and CWDEs accumulated in the tolerant rootstock (S21) grafts less than in susceptible scion (Rf) grafts. These results demonstrate that high levels of EPS and CWDE accumulation can be the cause of graft susceptibility upon bacterial wilt infection.

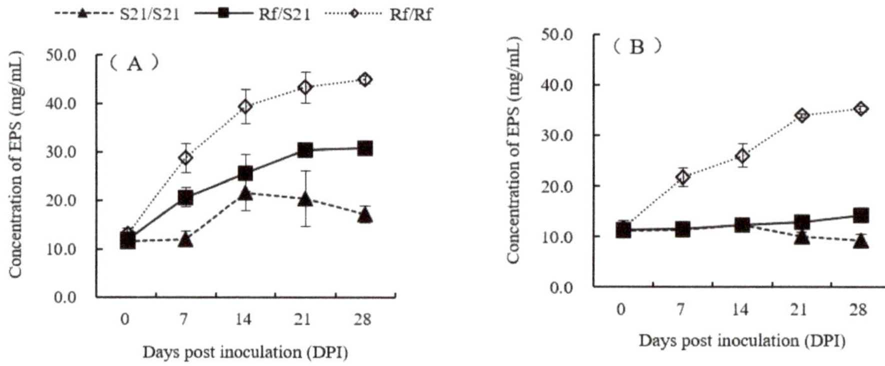

Figure 5. Concentration of the exopolysaccharides (EPSs) in the different graft combinations post-inoculation with *R. solanacearum*. Note (**A**) rootstock stems; (**B**) rootstock roots; the error bar indicates the standard error, *n* = 30. See Supplement for mean separations.

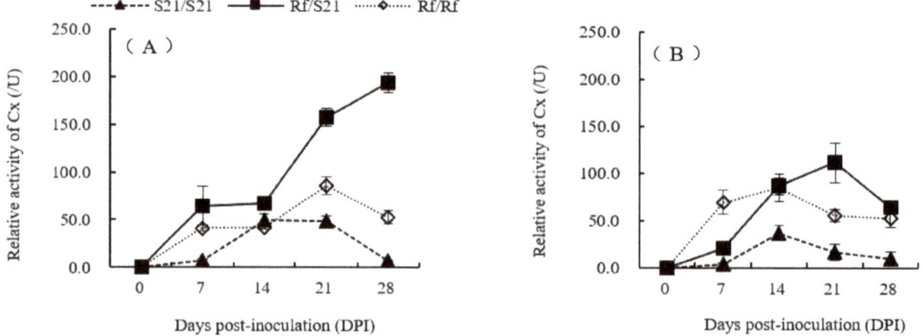

Figure 6. Relative activity of cellulase (Cx) in different graft combinations post-inoculation with *R. solanacearum*. Note (**A**) rootstock stems; (**B**) rootstock roots; the error bar indicates the standard error, *n* = 30. See Supplement for mean separations.

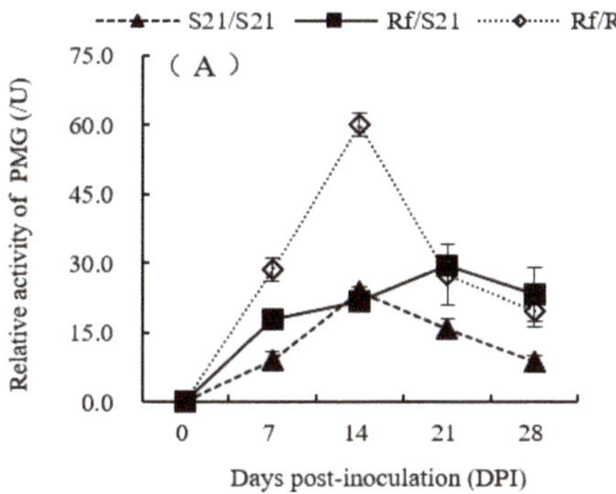
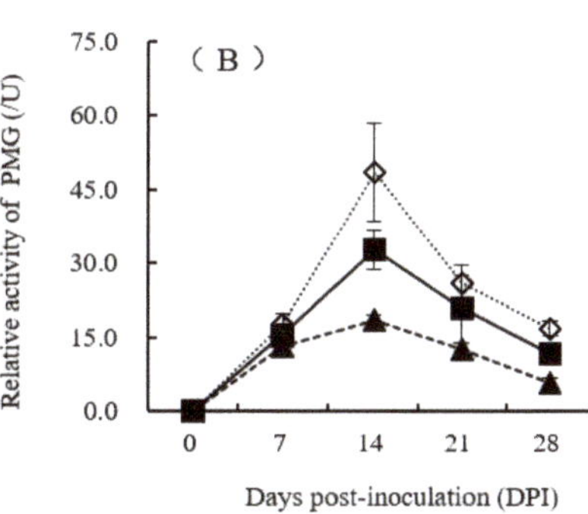

Figure 7. Relative activity of the pectin methylgalacturonase (PMG) in different graft combinations post-inoculation with *R. solanacearum*. Note (**A**) rootstock stems; (**B**) rootstock roots; the error bar indicates the standard error, n = 30. See Supplement for mean separations.

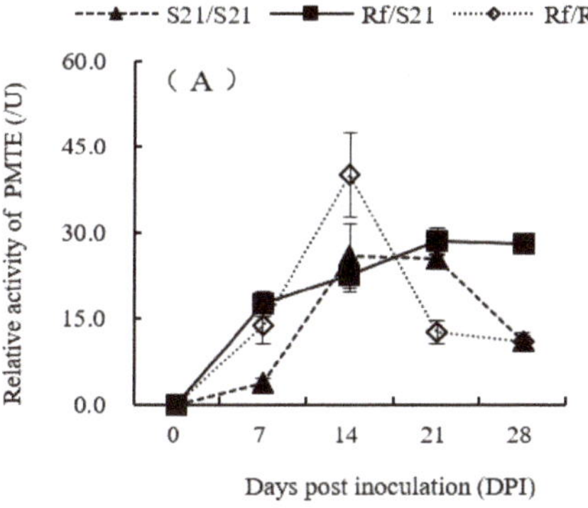
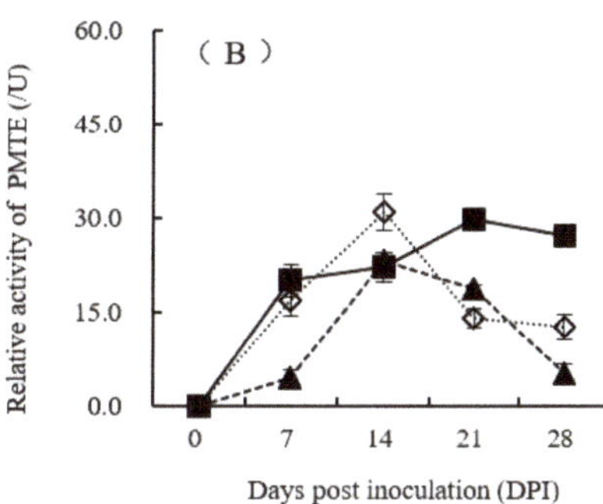

Figure 8. Relative activity of the pectin methyl transelimination enzyme (PMTE) in different graft combinations post-inoculation with *R. solanacearum*. Note (**A**) rootstock stems; (**B**) rootstock roots; the error bar indicates the standard error, n = 30. See Supplement for mean separations.

3.4. High Levels of ROS and Active Oxygen Scavenging Enzymes Accumulated in Susceptible Scion Grafts

In order to further investigate the defense factors of the tolerant rootstock against *R. solanacearum*, the level of ROS (content of O_2^- and H_2O_2) and the activities of active oxygen scavenging enzymes (SOD, CAD, POD and APX) were measured in the rootstocks (Data S3). The O_2^- and H_2O_2 content in each part of the S21/S21 rootstock were basically maintained at the same low levels, even at different DPIs with *R. solanacearum*, indicating that the ROS maintained their balance in the self-grafted rootstocks (Figures 9 and 10). The Rf/S21 rootstock reached a peak value for O_2^- concentration at 7 DPI with *R. solanacearum*, then this value began to decrease and ultimately became close to the O_2^- concentration of the S21/S21 rootstock at 14 DPI (Figure 9). In addition, the H_2O_2 content in each part of the Rf/S21 rootstock was always higher than that in S21/S21 (Figure 10). These results indicate that the susceptible scion (Rf) induced an increased accumulation of ROS in the tolerant rootstock (S21) upon *R. solanacearum* infection. Meanwhile, a high-level accumulation of ROS suggests that the susceptible scion (Rf) reduced tolerance of the rootstock (S21) causing *R. solanacearum* to more easily infect the host plant.

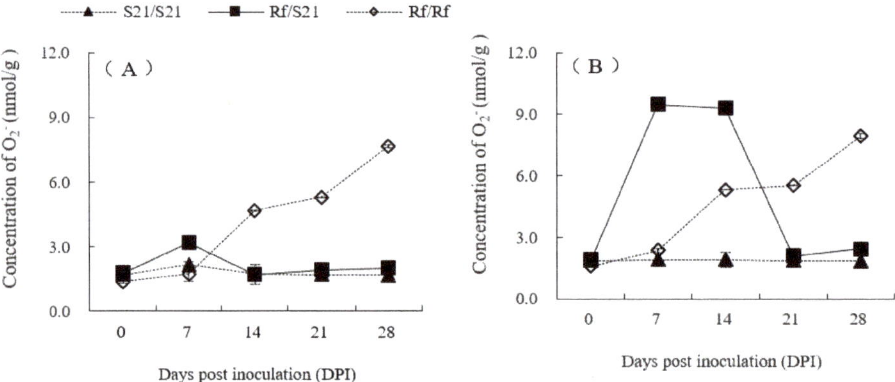

Figure 9. Concentration of O_2^- in different graft combinations post-inoculation with *R. solanacearum*. Note: (**A**): rootstock stems; (**B**): rootstock roots; the error bar indicates the standard error, $n = 30$. See Supplement for mean separations.

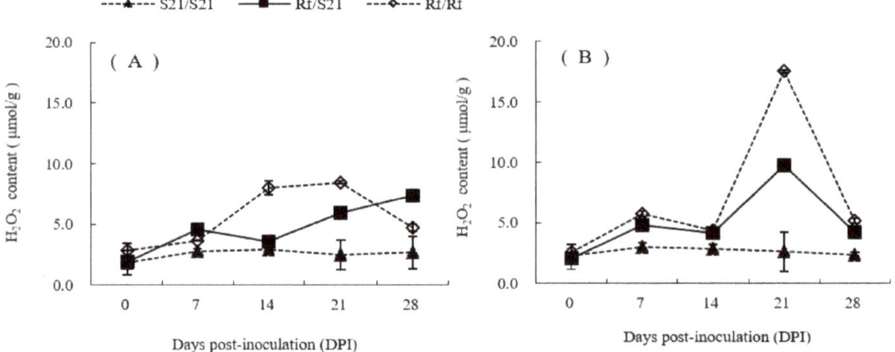

Figure 10. The H_2O_2 content in different graft combinations post-inoculation with *R. solanacearum*. Note: (**A**): rootstock stems; (**B**): rootstock roots; the error bar indicates the standard error, $n = 30$. See Supplement for mean separations.

In order to alleviate the damage of ROS on plant tissues, plants maintain a relative balance of ROS through active oxygen scavenging enzymes because more ROS is generated in response to pathogen infection. In addition to the SOD activity of the Rf/S21 rootstock (similar to S21/S21), the activity of CAD, POD, and APX were significantly higher in the Rf/S21 rootstock when compared with the S21/S21 rootstock post *R. solanacearum* inoculation (Figure 11). This also suggests that tolerance of the rootstock is affected by the susceptible scion, breaking the balance of the active oxygen scavenging system in the grafted rootstocks by the excess accumulation of ROS and active oxygen scavenging enzymes. The tolerant rootstock damages its own tissues via the imbalance of the active oxygen scavenging system when facing *R. solanacearum* infection in Rf/S21, resulting in reduced growth and further reducing its disease tolerance (Figures 2 and 4). Therefore, continuous infection of *R. solanacearum* in grafted eggplant would induce death in some of the grafted plants, even if they have a high tolerance in their rootstocks (Table 1).

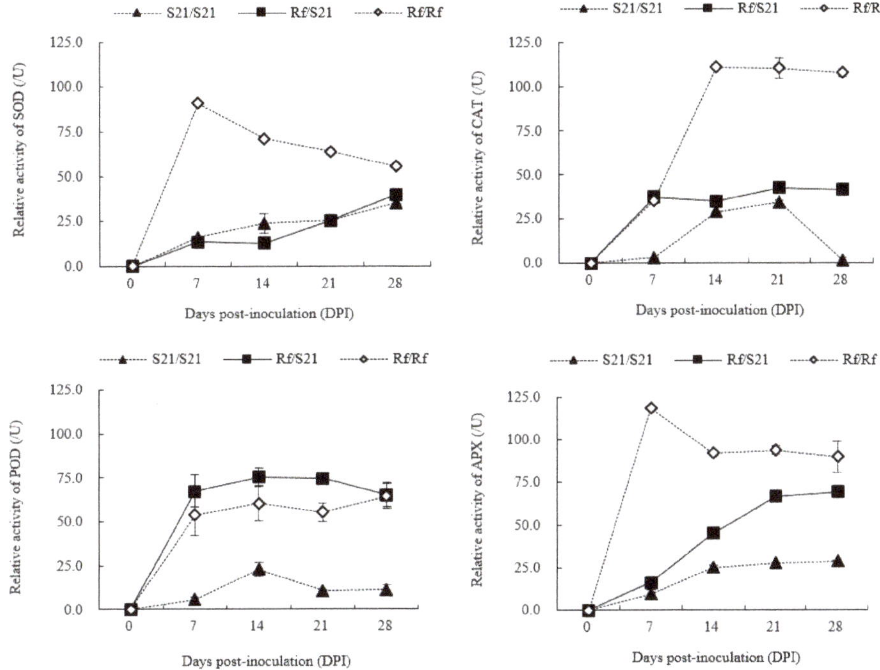

Figure 11. Relative activity of superoxide dismutase (SOD), catalase (CAT), peroxidase (POD), and ascorbate peroxidase (APX) of the rootstock roots in different graft combinations post-inoculation with *R. solanacearum*; the error bar indicates the standard error, *n* = 30. See Supplement for mean separations.

4. Discussion

A susceptible scion (Rf) on a tolerant rootstock reduced the biological yield of plants to a certain extent when infected with *R. solanacearum*, which corroborates results previously described for other plant species, including pepper (*Capsicum anuum L.*) and tomato (*Solanum lycopersicum*) [40,41]. Now, this phenomenon was documented for eggplant through this study. Currently, more attention has been given to the selection of a tolerant rootstock for grafting, while less attention has been given to the importance of scion tolerance [24]. The effect of the scion upon rootstock tolerance is significant in practical plant production, because susceptible scions cannot achieve their expected economic output due to a reduction in rootstock disease tolerance. Therefore, it will be meaningful to pay attention to the selection of disease tolerant scions in future vegetable grafting processes. The accumulation of EPS and the activity of CWDEs in eggplant rootstocks directly reflect the ability of rootstock plants to successfully react against *R. solanacearum* infection. Although it is not completely understood how these substances affect disease tolerance in grafted plants, the present results showed that *R. solanacearum* secreted massive amounts of EPS and CWDEs in rootstocks at 14 DPI with *R. solanacearum*. Pathogens could be prevented from harming eggplant growth if the inspection and defense work of plants is performed in time (before 14 DPI). Additionally, perhaps more effective EPS and CWDE detection systems could be developed to detect the infection of *R. solanacearum* in time and reduce losses as possible.

In plants, cell-to-cell communication plays a critical role in development, disease tolerance, and responses to diverse environmental stresses. Various types of plant RNA species move from cell to cell (short-range) or systemically (long-range) to potentially regulate whole-plant physiological processes [2,42]. Plant vascular systems are constructed by specific cell wall modifications, through which cells are highly specialized to make conduits for water and nutrients. *R. solanacearum* primarily causes pathogenesis by invading the vascular system, which obviously is involved in systemic substance circulation and regulation in plants. Plants resist the invasion of *R. solanacearum*, not only due to local tissue reactions, but also due to common systemic regulation. In this study susceptible scions showed a weak disease tolerance. It remains possible that some factors from susceptible scions move into tolerant rootstocks and reduce their tolerance. Recent studies have revealed that CLE and CEP peptides secreted in the roots are transported above ground via the xylem in response to plant–microbe interactions and soil nitrogen starvation, respectively [43]. Thus, plant vascular systems may not only act as conduits for the translocation of essential substances but also as long-distance communication pathways that allow plants to adapt to changes in internal and external environments at the whole plant level [44]. Our research on how susceptible scions influence the tolerance level of rootstocks has led us to propose a process involving the regulation of long-distance signals in plants. There may be a specific interaction between the scion and rootstock, and a grafted plant always maintains the exchange of substances through the xylem and phloem [45], in which disease-related substances (some toxic proteins) in the rootstock are transmitted upwards. If some substances (defense factors) from the scion are conducted downwards over time, the rootstock suffers weakened disease tolerance [46,47].

Although our experiment does not study the exchange or circulation of substances in grafted plants, the direct factors for the weakened disease tolerance in the tolerant rootstock were detected. These factors are directly associated with a decrease in disease tolerance, shown by the accumulation of EPSs and CWDEs in graft tissues (Figures 5–8). Furthermore, through measuring the related balance system of the ROS and active oxygen scavenging enzymes, the reasons for weakened disease tolerance of tolerant rootstocks were explained. The overproduction of O_2^- and H_2O_2 would damage plant tissues, while a complex dynamic balance in plants could offset these damages through the actions of SOD, CAT, POD, and APX (Figures 9–11). It is evident that proteins and a range of RNAs can be transported via the phloem, and some of these elements can exert their physiological functions in different plant tissues [44]. Therefore, it could be some unknown substance(s) secreted in the scion, and transport to the rootstock affecting its disease tolerance. In the future, the factors that regulate rootstock tolerance could be studied by big data analysis of RNA-seq to further study the mechanism of the interaction between scion and rootstock.

5. Conclusions

A susceptible scion reduced the disease tolerance of a bacterial wilt-tolerant rootstock upon *R. solanacearum* infection. Thus, the susceptible scions showed poorer growth. Moreover, the excess accumulation of EPS and CWDEs in plants with susceptible scions resulted in weakened disease tolerance of the rootstock. Furthermore, the excess accumulation of the ROS and active oxygen scavenging system disrupted the balance of the active oxygen scavenging system in the grafted plants, attenuating the disease tolerance of the rootstock.

Supplementary Materials: The following are available online at http://www.mdpi.com/2311-7524/5/4/78/s1, Table S1: Experiment arrangement, Data S1: Plant biomass, Data S2: EPS and CWDE analysis, Data S3: ROS related analysis.

Author Contributions: Conceptualization, C.H. and Y.W.; Methodology, Y.Y.; Software, Y.W.; Validation, C.H., Y.W. and Y.Y.; Formal analysis, Y.W.; Investigation, W.Y.; Resources, C.Z.; Data curation, M.N. and C.Z.; Writing—original draft preparation, C.H.; Writing—review and editing, C.H.; Visualization, M.N. and W.Y.; Supervision, W.Y.; Project administration, W.Y.; Funding acquisition, W.Y.

Funding: This work was supported by the National Nature Science Foundation of China (No. 31660568) and the Science and Technology Major Project of Guangxi (No. AA17204039-2, No. AA17204026-1).

Conflicts of Interest: The authors declare there are no conflict of interest regarding the publication of this manuscript. The funders had no role in the design of the study; in the collection, analyses, or interpretation of data; in the writing of the manuscript, or in the decision to publish the results.

References

1. Xu, J.Y.; Zhang, H.D.; Zhang, H.; Tong, Y.H.; Xu, Y.; Chen, X.J.; Ji, Z.L. Toxin produced by Rhizoctonia solani and its relationship with pathogenicity of the fungus. *J.-Yangzhou Univ. Agric. Life Sci. Ed.* **2004**, *25*, 61–64.
2. Zhang, X.Y.; Xu, K. Effect of interaction between rootstock and scion on chilling tolerance of grafted eggplant seedlings under low temperature and light conditions. *Sci. Agric. Sin.* **2009**, *42*, 3734–3740.
3. Tsuchiya, K. Molecular biological studies of Ralstonia solanacearum and related plant pathogenic bacteria. *Jpn. J. Phytopathol.* **2004**, *70*, 156–158. [CrossRef]
4. Angot, A.; Peeters, N.; Lechner, E.; Vailleau, F.; Baud, C.; Gentzbittel, L.; Sartorel, E.; Genschik, P.; Boucher, C.; Genin, S. Ralstonia solanacearum requires F-box-like domain-containing type III effectors to promote disease on several host plants. *Proc. Natl. Acad. Sci. USA* **2006**, *103*, 14620–14625. [CrossRef] [PubMed]
5. Wang, J.; Zhao, X.H.; Sun, S. Association between differentiation in pathogenicity and propagation & EPS production of two Ralstonia solanacearum isolates from eucalyptus. *For. Pest Dis.* **2008**, *3*, 4–6.
6. Leigh, J.A.; Coplin, D.L. Exopolysaccharides in plant-bacterial interactions. *Annu. Rev. Microbiol.* **1992**, *46*, 307–346. [CrossRef]
7. Yang, H.T.; Ren, X.Z.; Wang, S.J.; Wang, J.P.; Gao, Q. Study on distribution of antagonistic bacteria in root–soil system of tomato and its relation to occurrence of bacterial wilt. *Chin. J. Biol. Control* **1994**, *10*, 162–165.
8. Denny, T. Plant pathogenic Ralstonia species. In *Plant-Associated Bacteria*; Springer: Dordrecht, The Netherlands, 2007; pp. 573–644.
9. Liu, F.Z.; Lian, Y.; Feng, D.X.; Song, Y.; Chen, Y.H. Identification and Evaluation of Resistance to Bacterial Wilt in Eggplant Germplasm Resources. *J. Plant Genet. Resour.* **2005**, *6*, 381–385.
10. Zhang, S.; Sun, L.; Kragler, F. The phloem-delivered RNA pool contains small noncoding RNAs and interferes with translation. *Plant Physiol.* **2009**, *150*, 378–387. [CrossRef]
11. Cooper, R.M. The role of cell wall-degrading: Enzymes in infection and damage. In *Plant Disease: Infection, Damage and Loss*; Blackwell: Oxford, UK, 1984; pp. 13–27.
12. Wanjiru, W.M.; Zhensheng, K.; Buchenauer, H. Importance of cell wall degrading enzymes produced by Fusarium graminearum during infection of wheat heads. *Eur. J. Plant Pathol.* **2002**, *108*, 803–810. [CrossRef]
13. Mittler, R.; Vanderauwera, S.; Gollery, M.; Van Breusegem, F. Reactive oxygen gene network of plants. *Trends Plant Sci.* **2004**, *9*, 490–498. [CrossRef] [PubMed]
14. Torres, M.A. ROS in biotic interactions. *Physiol. Plant.* **2010**, *138*, 414–429. [CrossRef] [PubMed]
15. Liu, X.; Williams, C.E.; Nemacheck, J.A.; Wang, H.; Subramanyam, S.; Zheng, C.; Chen, M.S. Reactive oxygen species are involved in plant defense against a gall midge. *Plant Physiol.* **2010**, *152*, 985–999. [CrossRef] [PubMed]
16. Imlay, J.A. Pathways of oxidative damage. *Annu. Rev. Microbiol.* **2003**, *57*, 395–418. [CrossRef]
17. Baisak, R.; Rana, D.; Acharya, P.B.; Kar, M. Alterations in the activities of active oxygen scavenging enzymes of wheat leaves subjected to water stress. *Plant Cell Physiol.* **1994**, *35*, 489–495.
18. Ella, E.S.; Kawano, N.; Ito, O. Importance of active oxygen-scavenging system in the recovery of rice seedlings after submergence. *Plant Sci.* **2003**, *165*, 85–93. [CrossRef]
19. Cao, S.; Yang, Z.; Cai, Y.; Zheng, Y. Antioxidant enzymes and fatty acid composition as related to disease resistance in postharvest loquat fruit. *Food Chem.* **2014**, *163*, 92–96. [CrossRef]
20. Shah, K.; Kumar, R.G.; Verma, S.; Dubey, R.S. Effect of cadmium on lipid peroxidation, superoxide anion generation and activities of antioxidant enzymes in growing rice seedlings. *Plant Sci.* **2001**, *161*, 1135–1144. [CrossRef]
21. Yamaguchi, K.; Takahashi, Y.; Berberich, T.; Imai, A.; Miyazaki, A.; Takahashi, T.; Michael, A.; Kusano, T. The polyamine spermine protects against high salt stress in *Arabidopsis thaliana*. *FEBS Lett.* **2006**, *580*, 6783–6788. [CrossRef]
22. Duan, X.; Bi, H.G.; Li, T.; Wu, G.X.; Li, Q.M.; Ai, X.Z. Root characteristics of grafted peppers and their resistance to *Fusarium solani*. *Biol. Plant.* **2017**, *61*, 579–586. [CrossRef]
23. Zorov, D.B.; Juhaszova, M.; Sollott, S.J. Mitochondrial reactive oxygen species (ROS) and ROS-induced ROS release. *Physiol. Rev.* **2014**, *94*, 909–950. [CrossRef] [PubMed]
24. King, S.R.; Davis, A.R.; Liu, W.; Levi, A. Grafting for disease resistance. *HortScience* **2008**, *43*, 1673–1676. [CrossRef]

25. Louws, F.J.; Cary, L.R.; Chieri, K. Grafting fruiting vegetables to manage soilborne pathogens, foliar pathogens, arthropods and weeds. *Sci. Hortic.* **2010**, *127*, 127–146. [CrossRef]

26. Martínez-Ballesta, M.C.; Alcaraz-López, C.; Muries, B.; Mota-Cadenas, C.; Carvajal, M. Physiological aspects of rootstock–scion interactions. *Sci. Hortic.* **2010**, *127*, 112–118. [CrossRef]

27. Bletsos, F.A.; Christos, M. Olympios. Rootstocks and grafting of tomatoes, peppers and eggplants for soil-borne disease resistance, improved yield and quality. *Eur. J. Plant Sci. Biotechnol.* **2008**, *2*, 62–73.

28. Aloni, B.; Cohen, R.; Karni, L.; Aktas, H.; Edelstein, M. Hormonal signaling in rootstock–scion interactions. *Sci. Hortic.* **2010**, *127*, 119–126. [CrossRef]

29. Nakaho, K.; Inoue, H.; Takayama, T.; Miyagawa, H. Distribution and multiplication of Ralstonia solanacearum in tomato plants with resistance derived from different origins. *J. Gen. Plant Pathol.* **2004**, *70*, 115–119. [CrossRef]

30. Seemüller, E.; Harries, H. *Plant Resistance. Phytoplasmas: Genomes, Plant Hosts and Vectors*; CAB International: Oxfordshire, UK, 2010; pp. 147–169.

31. Johnson, S.J.; Kreider, P.; Miles, C.A. *Vegetable Grafting: Eggplants and Tomatoes*; Washington State University Extension: Pullman, WA, USA, 2011.

32. Ramesh, R.; Joshi, A.A.; Ghanekar, M.P. Pseudomonads: Major antagonistic endophytic bacteria to suppress bacterial wilt pathogen, Ralstonia solanacearum in the eggplant (*Solanum melongena* L.). *World J. Microbiol. Biotechnol.* **2009**, *25*, 47–55. [CrossRef]

33. Liu, H.; Zhang, S.; Schell, M.A.; Denny, T.P. Pyramiding unmarked deletions in Ralstonia solanacearum shows that secreted proteins in addition to plant cell-wall-degrading enzymes contribute to virulence. *Mol. Plant-Microbe Interact.* **2005**, *18*, 1296–1305. [CrossRef]

34. Cao, J.K.; Jiang, W.B.; Zhao, Y.M. *Experiment Guidance of Postharvest Physiology and Biochemistry of Fruits and Vegetables*; China Light Industry Press: Beijing, China, 2007; pp. 84–87.

35. Li, B.; Zhou, C.; Zhao, K.; Li, F.; Chen, H. Pathogenic mechanism of scab of cucumber caused by Cladosporium cucumerinum II: The cell wall-degrading enzymes and its pathogenic action. *Acta Phytopathol. Sin.* **2000**, *30*, 13–18.

36. Yoshie, Y.; Goto, K.; Takai, R.; Iwano, M.; Takayama, S.; Isogai, A.; Che, F.S. Function of the rice gp91phox homologs OsrbohA and OsrbohE genes in ROS-dependent plant immune responses. *Plant Biotechnol.* **2005**, *22*, 127–135. [CrossRef]

37. Dhindsa, R.S.; Plumb-Dhindsa, P.; Thorpe, T.A. Leaf senescence: Correlated with increased levels of membrane permeability and lipid peroxidation, and decreased levels of superoxide dismutase and catalase. *J. Exp. Bot.* **1981**, *32*, 93–101. [CrossRef]

38. Kochba, J.; Lavee, S.; Spiegel-Roy, P. Differences in peroxidase activity and isoenzymes in embryogenic ane non-embryogenic 'Shamouti' orange ovular callus lines. *Plant Cell Physiol.* **1977**, *18*, 463–467. [CrossRef]

39. Dalton, D.A.; Hanus, F.J.; Russell, S.A.; Evans, H.J. Purification, properties, and distribution of ascorbate peroxidase in legume root nodules. *Plant Physiol.* **1987**, *83*, 789–794. [CrossRef] [PubMed]

40. Liu, Y.; Jiang, F.; Zhang, N.; Wang, H.; Ai, X. Relationship between osmoregulation and bacterial wilt resistance of grafted pepper. *Acta Hortic. Sin.* **2011**, *38*, 903–910.

41. McAvoy, T.; Freeman, J.H.; Rideout, S.L.; Olson, S.M.; Paret, M.L. Evaluation of grafting using hybrid rootstocks for management of bacterial wilt in field tomato production. *HortScience* **2012**, *47*, 621–625. [CrossRef]

42. Lewsey, M.G.; Hardcastle, T.J.; Melnyk, C.W.; Molnar, A.; Valli, A.; Urich, M.A.; Nery, J.R.; Baulcombe, D.C.; Ecker, J.R. Mobile small RNAs regulate genome-wide DNA methylation. *Proc. Natl. Acad. Sci. USA* **2016**, *113*, E801–E810. [CrossRef]

43. Oh, E.; Pil, J.S.; Jungmook, K. Signaling peptides and receptors coordinating plant root development. *Trends Plant Sci.* **2018**, *23*, 337–351. [CrossRef]

44. Notaguchi, M.; Satoru, O. Dynamics of long-distance signaling via plant vascular tissues. *Front. Plant Sci.* **2015**, *6*, 161. [CrossRef]

45. Noctor, G.; Gomez, L.; Vanacker, H.; Foyer, C.H. Interactions between biosynthesis, compartmentation and transport in the control of glutathione homeostasis and signaling. *J. Exp. Bot.* **2002**, *53*, 1283–1304. [CrossRef]

46. Jiang, F.; Liu, Y.; Ai, X.; Zheng, N.; Wang, H. Study on relationship among microorganism, enzymes' activity in rhizosphere soil and root rot resistance of grafted capsicum. *Sci. Agric. Sin.* **2010**, *43*, 3367–3374.

47. Chattopadhyay, C.; Kolte, S.J.; Waliyar, F. *Diseases of Edible Oilseed Crops*; CRC Press: Boca Raton, FL, USA, 2015.

 *horticulturae*

Article

Association of Indolebutyric Acid with *Azospirillum brasilense* in the Rooting of Herbaceous Blueberry Cuttings

Renata Koyama, Walter Aparecido Ribeiro Júnior, Douglas Mariani Zeffa, Ricardo Tadeu Faria, Henrique Mitsuharu Saito, Leandro Simões Azeredo Gonçalves and Sergio Ruffo Roberto *

Agricultural Research Center, Londrina State University, Celso Garcia Cid Road, km 380, P.O. Box 10.011, Londrina ZIP 86057-970, Brazil; emykoyama@hotmail.com (R.K.); junior_agro40@hotmail.com (W.A.R.J.); douglas.mz@hotmail.com (D.M.Z.); faria@uel.br (R.T.F.); henriquesaito.m@gmail.com (H.M.S.); leandrosag@uel.br (L.S.A.G.)
* Correspondence: sroberto@uel.br

Received: 5 July 2019; Accepted: 10 September 2019; Published: 1 October 2019

Abstract: Association between auxins and plant growth-promoting bacteria can stimulate root growth and development of fruit crop nursery plants, and can be a promising biological alternative to increase the rooting of cuttings. The objective of this study was to assess the viability of producing 'Powderblue' blueberry nursery plants from cuttings using different doses of indolebutyric acid (IBA) in association with *Azospirillum brasilense*. The following treatments were tested: 0 (control); 500 mg L^{-1} of IBA; 1000 mg L^{-1} of IBA; *A. brasilense*; 500 mg L^{-1} of IBA + *A. brasilense*; and 1000 mg L^{-1} of IBA + *A. brasilense*. The experimental design was completely randomized, with six treatments and four replicates, and each plot (box) consisted of 10 cuttings. The boxes were arranged in a mist chamber with an intermittent regimen controlled by a timer and solenoid valve. After 90 days, the following variables were assessed: rooted cuttings; survival of cuttings; foliar retention; sprouting; cuttings with callus; root dry mass per cutting; number of roots per cutting; and root length. It was observed that the application of IBA with the *A. brasilense* rhizobacteria increased the number of roots of 'Powderblue' blueberry cuttings, while the treatments with IBA alone and IBA 1000 mg L^{-1} + *A. brasilense* increased the root length of cuttings. However, treatments with IBA and *A. brasilense* had no impact on % rooted cuttings and % survival of cuttings.

Keywords: auxin; rhizobacteria; rooting; Vaccinium spp.

1. Introduction

Blueberry (*Vaccinium* spp.), a fruit species of temperate climates, has high medicinal value owing to its high content of anthocyanins which confer high antioxidant properties to the fruits. These antioxidants help neutralize free radicals related to the onset of degenerative diseases, such as cancer, cataracts, immune disorders, cognitive impairments, and muscle degeneration [1]. The commercial interest in this species has greatly increased because of several studies reporting its efficacy as a medicine [2–4]. Consequently, its cultivation is becoming widespread and of interest to producers in Brazil, where it was first grown around 1983 in Pelotas (RS), in studies conducted by the Brazilian Agricultural Research Corporation [5].

Blueberry is mostly asexually propagated through cuttings as adult plants are obtained in a shorter time period compared to those sexually propagated. In addition, this method can be easily executed and lead to the formation of adventitious roots in cuttings in any season of the year [6,7]. However, different blueberry cultivars show distinct responses regarding rooting of cuttings owing to genetic factors that influence cell differentiation and root formation [8,9].

For plant species that are difficult to root, such as blueberry, auxiliary techniques with growth regulators, such as indolebutyric acid (IBA), can be used [10]. Exogenous application of synthetic auxins is very common in vegetative propagation methods using cuttings as they act similar to the natural phytohormone [11]. However, studies have shown that the use of the synthetic hormone IBA in high concentrations can be toxic to the plant or inhibit rooting [12].

The use of plant growth-promoting rhizobacteria (PGPRs) can be a promising biological alternative for increasing the rooting of cuttings [13]. *Azospirillum* is a genus of PGPRs that inhabits the roots of host plants and provides beneficial effects to the plant under normal growth and/or stress conditions [14]. PGPRs of this genus can increase the fixation of free nitrogen and the production of phytohormones, thus promoting growth in inoculated plants [15].

Studies have shown promising results for association with the PGPR species *Azospirillum brasilense*; it promotes root development by increasing the production of hormones, leading to growth and development of plants [16–18]. In this context, the objective of this study was to assess the viability of producing blueberry nursery plants from cuttings using different doses of IBA in association with *A. brasilense*.

2. Materials and Methods

2.1. Plant Materials

The experiment was conducted from April to July 2018 at the Department of Horticulture of the Agricultural Research Center of the Londrina State University—PR, Brazil (latitude 23°23′ S, longitude 51°11′ W and elevation of 566 m). Herbaceous cuttings from the median part of shoots of 'Powderblue' (*Vaccinium* sp.) blueberry stock plants were used and maintained in the same department in a greenhouse, originating from a collection of blueberry cultivars belonging to the Brazilian Agricultural Research Corporation, Pelotas, RS, Brazil.

2.2. Experimental Design and Treatments Description

The experimental design was completely randomized, with six treatments and four replicates, for a total of 24 experimental units, with each plot comprised of 10 cuttings. The effect of different concentrations of IBA in talc and the association with *A. brasilense* was assessed. The following treatments were tested: 0 (control); 500 mg L^{-1} of IBA; 1000 mg L^{-1} of IBA; *A. brasilense*; 500 mg L^{-1} of IBA + *A. brasilense*; and 1000 mg L^{-1} of IBA + *A. brasilense*. The maximum tested dose of IBA (1000 mg L^{-1}) was based on Fischer et al. [19].

Before collecting the cuttings, the IBA solutions were prepared. For this, 0.05 and 0.1 g of concentrated IBA (99.9% purity; Sigma-Aldrich®, St. Louis, MO, USA) were weighed using a semi-analytical balance and dissolved in 50 mL of 100% ethanol. When the IBA was totally dissolved, the volume was adjusted to 100 mL with distilled water, and solutions with concentrations of 500 and 1000 mg L^{-1} of IBA were obtained. To prepare the IBA in talc, 0.1 g of IBA was mixed in industrial inert talc (Quimidrol®, Joinville, Brazil) for a total of 100 g. For better homogenization, sufficient IBA with ethanol solutions was added to form a paste, and then transferred to an oven at 40 °C, where it remained until complete evaporation of the solvent.

Cuttings were prepared using a bevel cut just below a node, discarding the leaves at the basal part, and keeping two pairs of leaves in the upper part which were later cut in half. During preparation, the cuttings were kept in a container with water to avoid dehydration. After preparing the cuttings, the IBA talc was applied to the base of the cuttings.

Then, the cuttings were placed in perforated plastic boxes (44 × 30 × 7 cm) containing vermiculite substrate of medium granulation for rooting, followed by the application of 1 mL of solution containing *A. brasilense* (BR 11001t) isolates obtained by Silva et al. [20]. The bacterial isolates were cultured in DYGS medium (glucose 2%, peptone 1.5%, yeast extract 2%, KH$_2$PO$_4$ 0.5%, MgSO$_4$·7H$_2$O 0.5%, glutamic acid 1.5%, pH 6.8), the procedure recommended by Kuss [21], with agitation of 120× *g* at

28 °C, until reaching a population density of 10^8 CFU mL^{-1}. The cultured bacteria were applied directly into the substrate.

2.3. Growth Conditions

The boxes were arranged in a mist chamber with an intermittent misting regimen controlled by a timer and solenoid valve. The valve was programmed to mist for 10 s every 6 min. The misting nozzle employed (Model DAN-7755 Modular Greenhouse Sprinkler, Tel Aviv, Israel) had a flow rate of 35 L h^{-1}. The mist chamber was placed in a greenhouse with a transparent polyethylene film and 30% cover.

To control fungal diseases, the cuttings were treated weekly with a spray of tebuconazole fungicides (1 mL L^{-1}). Foliar fertilization with Biofert Plus® (Contagem, Brazil) fertilizer (8-9-9 + micronutrients) at a 5 mL L^{-1} concentration per spray was applied every 15 d.

2.4. Evaluations

After 90 d, the following variables were assessed: rooted cuttings (% of cuttings that developed at least one root); survival of cuttings (% of live cuttings); foliar retention (% of cuttings that did not lose leaves); sprouting (shoot growth); cuttings with callus (% of live cuttings without roots); root dry mass per cutting (g); number of roots per cutting; and root length (cm). The root dry mass was determined by oven drying samples with forced air circulation at 78 °C for 24 h.

2.5. Statistical Analysis

Prior to individual analysis of variance, the homoscedasticity and normality of the residues were confirmed using Bartlett [22] ($p \leq 0.05$) and Lilliefors [23] tests ($p \leq 0.05$), respectively. Arc-sine transformations $\sqrt{(x/100)}$ were performed for variables expressed in percentage and the transformation $\sqrt{(x + 1)}$ was performed for counting data. After performing the individual analysis of variance, those characteristics which had fewer than seven ratios, between the highest and the lowest residuals, were submitted to factorial analysis of variance. Means of significantly affected characteristics were compared by Tukey's test ($p \leq 0.05$).

3. Results and Discussion

There were no significant differences for % rooted cuttings, survival, foliar retention, cuttings with callus, and root dry mass (Table 1). In contrast to our results, olive-tree cuttings (*Olea europaea*) treated with 3 g L^{-1} of IBA in association with *A. brasilense* led to a higher percentage of rooted cuttings [24]. Therefore, it is likely that the application of IBA and the inoculation with rhizobacteria present different responses depending on the species.

Table 1. Rooted cuttings (RC), survival (SV), foliar retention (FR), leaf retention (LR), sprouting (S), cuttings with callus (CC), root dry mass (DM), number of roots (NR), and root length (RL) of 'Powderblue' blueberry cuttings exposed to indolebutyric acid and *Azospirillum brasilense*.

Sources of Variation	RC (%)	SV (%)	FR (%)	S (%)	CC (%)	DM (g)	NR	RL (cm)
TMS [z]	0.1 [ns] [y]	0.3 [ns]	0.4 [ns]	0.8 *	0.4 [ns]	5.5 [ns]	27.3 *	66.5 *
EMS	0.3	0.2	0.4	0.2	0.3	4.1	8.5	11.4
Mean [x]	45.4	89.2	42.0	73.3	37.1	0.6	3.9	4.5
CV (%)	16.7	8.8	12.3	12.2	20.8	33.8	15.0	26.6

[z] TMS = Treatment mean square. EMS = Error mean square. [y] * = significant $p \leq 0.05$ by ANOVA. ns: not significant. [x] Mean across treatments. CV (%) = Coefficient of variation.

The applications of IBA and the rhizobacterium *A. brasilense* increased the number of roots per cutting (Tables 1 and 2). One of the great advantages of increased lateral root production through cuttings is that the volume of soil available to the plant is higher, indirectly increasing the capture of available resources in the soil and increasing the absorption of water and nutrients. These results were also observed with 'Berkeley' blueberry from the application of 1000 mg L^{-1} IBA [25] and with 'Bluegem' and 'Powderblue' blueberries from the application of 2000 mg L^{-1} IBA [8].

Table 2. Number of roots (NR), root length (RL), and sprouting (S) of 'Powderblue' blueberry cuttings treated with indolebutyric acid (IBA) and *Azospirillum brasilense* (Azo).

Treatments	NR [1/]	RL (cm) [1/]	S (%) [1/]
Control	2.79 c	2.81 b	92.5 a
500 mg L^{-1} IBA	4.02 b	5.62 a	85.0 a
1000 mg L^{-1} IBA	3.50 b	5.18 a	60.0 b
Azo	5.57 a	3.88 ab	70.0 b
Azo + 500 mg L^{-1} IBA	4.38 b	4.28 ab	75.0 b
Azo + 1000 mg L^{-1} IBA	3.38 b	5.00 a	57.5 c

[1/] Means followed by the same letter in a column do not differ by the Tukey's test set at 5% probability.

In addition, it was possible to observe that inoculation of *A. brasilense* alone was superior to the other treatments in increasing the number of roots per cutting. *A. brasilense* is known for the ability to alter the architecture of the roots of plants, increasing the formation of lateral and adventitious roots and root hairs. This is because of the production or metabolization by *A. brasilense* of chemical signaling compounds that alter root elongation and arrangement and the formation of root hairs [26].

However, there were no significant differences between treatments with IBA alone and in association with *A. brasilense*. Therefore, such an association did not benefit the development of new roots compared to the treatments with IBA alone. Indolacetic acid (IAA) is one of the most important molecules produced by *Azospirillum* sp. and studies suggest bacterial biosynthesis of IAA can be drastically affected by environmental conditions and exposure to other compounds [16,18]. Thus, the application of IBA may have influenced the production of the rhizobacteria growth-promoting substances.

The mean root length (Table 2) increased with applications of 500 and 1000 mg L^{-1} of IBA and 1000 mg L^{-1} of IBA + *A. brasilense,* and cuttings showed more developed roots compared to that of the control. Genetic factors can influence cell differentiation and root formation in cuttings, and different blueberry cultivars show distinct responses regarding the rooting of cuttings. In studies conducted with 'Climax' and 'Florida' blueberries, the application of IBA did not increase root length of the roots [27], but 'Bluegem', 'Bluebelle', and 'Powderblue' blueberries responded exponentially to the different concentrations of the growth regulator [8].

Regarding sprouting, emergence of growing buds was more vigorous in the control as compared to treatment with 500 mg L^{-1} of IBA. Initiation and development of buds are independent of the adventitious formation of roots. The application of high concentrations of auxins in cuttings can inhibit bud growth, sometimes even preventing growth of the aerial part, although root formation may be adequate [11]. Therefore, it was verified that application of *A. brasilense* alone is a viable alternative to increasing the number of adventitious roots in 'Powderblue' blueberry cuttings.

4. Conclusions

The application of IBA with the *Azospirillum brasilense* rhizobacteria increased the number of roots of 'Powderblue' blueberry cuttings, while the treatments with IBA alone and IBA 1000 mg L^{-1} + *A. brasilense* increased the root length of cuttings. However, treatments with IBA and *A. brasilense* combined had no impact on % rooted cuttings and % survival of cuttings.

Author Contributions: R.K. and D.M.Z. conceived and designed the experiments. R.K. and H.M.S. performed the experiments. R.K., W.A.R.J., R.T.F., L.S.A.G. and S.R.R. wrote the manuscript. D.M.Z. analyzed the data.

Funding: This study was funded by the Brazilian Council for Scientific and Technological Development (CNPq) and Coordination for the Improvement of Higher Education Personnel (CAPES).

Conflicts of Interest: The authors declare no conflicts of interest.

References

1. Retamales, J.B.; Hancock, J.F. *Blueberries*, 1st ed.; CABI: Wallingford, UK, 2012; pp. 19–50.
2. Shi, M.; Loftus, H.; Mcainch, A.J.; Su, X.Q. Blueberry as a source of bioactive compounds for the treatment of obesity, type 2 diabetes and chronic inflammation. *J. Funct. Foods* **2017**, *30*, 16–29. [CrossRef]
3. Yan, Y.; Zhang, F.; Chai, Z.; Liu, M.; Battino, M.; Meng, X. Mixed fermentation of blueberry pomace with L. rhamnosus GG and L. plantarum-1: Enhance the active ingredient, antioxidant activity and health-promoting benefits. *Food Chem. Toxicol.* **2019**, *131*, 110541. [CrossRef]
4. Wang, Y.; Fong, S.K.; Singh, A.P.; Vorsa, N.; Cicalese, N.J. Variation of anthocyanins, proanthocyanidins, flavonols, and organic acids in cultivated and wild diploid blueberry species. *HortScience* **2019**, *54*, 576–585. [CrossRef]
5. Antunes, L.E.C.; Raseira, M.C.B. *Cultivo do mirtilo (Vaccinium spp.)*, 1st ed.; EMBRAPA: Pelotas, Brazil, 2006; pp. 1–98.
6. Shahab, M.; Roberto, S.R.R.; Colombo, R.C.; Silvestre, J.P.; Ahmed, S.; Koyama, R.; Hussain, I. Clonal propagation of blueberries minicuttings under subtropical conditions. *Int. J. Biosci.* **2018**, *13*, 1–9. [CrossRef]
7. Porfírio, S.; Silva, M.D.R.; Cabrita, M.J.; Azadi, P.; Peixe, A. Reviewing current knowledge on olive (*Olea europaea* L.) adventitious root formation. *Sci. Hort.* **2016**, *198*, 207–226. [CrossRef]
8. Marangon, M.A.; Biasi, L.A. Cutting propagation of blueberry in seasons of the year with indolebutyric acid and bottom heat. *Pesq. Agropec. Bras.* **2013**, *48*, 25–32. [CrossRef]
9. Pio, R.; Costa, F.C.; Curi, P.N.; Moura, P.H.A. Rooting of kiwi cultivars woody cuttings. *Sci. Agric.* **2010**, *11*, 271–274. Available online: https://www.redalyc.org/html/995/99515056012/ (accessed on 24 June 2019).
10. Vignolo, G.K.; Fischer, D.L.O.; Araujo, V.F.; Kunde, R.J.; Antunes, L.E.C. Rooting of hardwood cuttings of three blueberry cultivars with different concentrations of IBA. *Cienc. Rural* **2012**, *42*, 795–800. [CrossRef]
11. Hartmann, H.T.; Kester, D.E.; Davies, F.T.; Geneve, R.L. *Plant Propagation: Principles and Practices*, 8th ed.; Pearson: Boston, MA, USA, 2011; pp. 295–360.
12. Sauer, M.; Robert, S.; Kleine-Vehn, J. Auxin: Simply complicated. *J. Exp. Bot.* **2013**, *64*, 2565–2577. [CrossRef]
13. Mariosa, T.N.; Melloni, E.G.P.; Melloni, R.; Ferreira, G.M.R.; Souza, S.M.P.; Silva, L.F.O. Rhizobacteria and development of seedlings from semi-hardwood cuttings of olive (*Olea europaea* L.). *Rev. Cienc. Agrar.* **2017**, *60*, 302–306. [CrossRef]
14. Khademian, R.; Asghari, B.; Sedaghati, B.; Yaghoubian, Y. Plant beneficial rhizospheric microorganisms (PBRMs) mitigate deleterious effects of salinity in sesame (*Sesamum indicum* L.): Physio-biochemical properties, fatty acids composition and secondary metabolites content. *Ind. Crops. Prod.* **2019**, *136*, 129–139. [CrossRef]
15. Cassán, F.; Diaz-Zorita, M. *Azospirillum* sp. in current agriculture: From the laboratory to the field. *Soil Biol. Biochem.* **2016**, *103*, 117–130. [CrossRef]
16. Molina, R.; Rivera, D.; Mora, V.; López, G.; Rosas, S.; Spaepen, S.; Vanderleyden, J.; Cassán, F. Regulation of IAA biosynthesis in *Azospirillum brasilense* under environmental stress conditions. *Curr. Microbiol.* **2018**, *75*, 1408–1418. [CrossRef] [PubMed]
17. Rivera, D.; Mora, V.; Lopez, G.; Rosas, S.; Spaepen, S.; Vanderleyden, J.; Cassan, F. New insights into indole-3-acetic acid metabolism in *Azospirillum brasilense*. *J. Appl. Microbiol. Biochem.* **2018**, *125*, 1774–1785. [CrossRef]
18. Fukami, J.; Ollero, F.J.; de la Osa, C.; Valderrama-Fernández, R.; Nogueira, M.A.; Megías, M.; Hungria, M. Antioxidant activity and induction of mechanisms of resistance to stresses related to the inoculation with *Azospirillum brasilense*. *Arch. Microbiol.* **2018**, 1191–1203. [CrossRef]
19. Fischer, D.L.O.; Fachinello, J.C.; Antunes, L.E.C.; Tomaz, Z.F.P.; Giacobbo, C.L. Efeito do ácido indolbutírico e da cultivar no enraizamento de estacas lenhosas de mirtilo. *Rev. Bras. Frutic.* **2008**, *30*, 285–289. [CrossRef]
20. Silva, T.F.; Melloni, R. Density and phenotypic diversity of non-symbiotic diazotrophic bacteria in soils of the Biological Reserve Serra dos Toledos, Itajubá (MG). *Rev. Bras. Ciênc.* **2011**, *35*, 359–371. [CrossRef]
21. Kuss, A.V.; Kuss, V.V.; Lovato, T.; Flores, M.L. Nitrogen fixation and in vitro production of indolacetic acid by endophytic diazotrophic bacteria. *Pesq. Agropec. Bras.* **2007**, *42*, 1459–1465. [CrossRef]

22. Bartlett, M.S. Properties of Sufficiency and Statistical Tests. Proceedings of the Royal Society of London. *J. Math. Phys. Sci.* **1937**, *160*, 268–282. [CrossRef]
23. Lilliefors, H.W. On the Kolmogorov-Smirnov test for normality. *Biometrika* **1967**, *62*, 399–402. [CrossRef]
24. Rosa, D.D.; Villa, F.; Silva, D.F.; Corbari, F. Rooting of semihardwood cuttings of olive: Indolbutyric acid, calcium and *Azospirillum brasilense*. *Comun. Sci.* **2018**, *9*, 34–40. Available online: https://www.comunicatascientiae.com.br/comunicata/article/view/977/528 (accessed on 24 June 2019). [CrossRef]
25. Çelik, H. The performance of some northern highbush blueberry (*Vaccinium corymbosum* L.) varieties in north eastern part of Anatolia. *Anadolu Tarım Bilim. Derg.* **2009**, *24*, 141–146. Available online: http://www3.omu.edu.tr/anajas/pdf/24(3)/141-146.pdf (accessed on 24 June 2019).
26. Gouda, S.; Kerry, R.J.; Samal, D.; Mahapatra, G.P.; Das, G.; Patra, J.K. Application of plant growth promoting rhizobacteria in agriculture. In *Advances in Microbial Biotechnology*, 1st ed.; Kumar, P., Patra, J.K., Chandra, J., Eds.; Apple Academic Press: New York, NY, USA, 2018; pp. 73–83.
27. Peña, M.L.P.; Gubert, C.; Tagliani, M.C.; Bueno, P.M.C.; Biasi, L.A. Concentrations and forms of application of indolebutyric acid on cutting propagation of cvs. Florida and Climax blueberries. *Semina* **2012**, *33*, 57–64. [CrossRef]

 *horticulturae*

Article

In Vitro Propagation and Acclimatization of Dragon Tree (*Dracaena draco*)

Alexis Galus [1], Ali Chenari Bouket [2] and Lassaad Belbahri [1,3,*]

[1] Nextbiotech, 98 Rue Ali Belhouane, Agareb 3030, Tunisia
[2] Plant Protection Research Department, East Azarbaijan Agricultural and Natural Resources Research and Education Center, AREEO, Tabriz 5355179854, Iran
[3] Laboratory of Soil Biology, University of Neuchatel, 2000 Neuchatel, Switzerland
* Correspondence: lassaad.belbahri@unine.ch; Tel.: +41-76-576-3069

Received: 24 June 2019; Accepted: 29 August 2019; Published: 6 September 2019

Abstract: In this study, an efficient in vitro procedure was developed for bud induction, rooting of developing shoots and greenhouse acclimatization of young plantlets of dragon tree (*Dracaena draco*). Effects of media (S1 (1 mg/L KIN and 1 mg/L NAA), S2 (3 mg/L KIN and 1 mg/L IAA), S3 (1 mg/L BAP and 2 mg/L IBA) and S4 (1 mg/L BAP and 1 mg/L NAA)) on shoot induction and media (R1 (0 mg/L IBA), R2 (0.5 mg/L IBA), R3 (1 mg/L IBA), and R4 (2 mg/L IBA)) on root induction were examined in order to find optimal plant hormone concentrations for efficient *Dracaena draco* dormant bud development and subsequent rooting. The best shoot induction and rooting media were S1 and S2, and R3 and R4, respectively. Dormant buds from one-year-old *Dracaena draco* plants submitted to this in vitro procedure allowed successful recovery of up to 8 individuals per explant used. In vitro grown plants were successfully acclimated in the greenhouse. The potential of this in vitro procedure for multiplication of this endangered tree is discussed in this report.

Keywords: acclimatization; auxins; cytokinins; *Dracaena draco*; in vitro

---

1. Introduction

Dracaena draco, the Dragon Tree or Drago, is a subtropical plant native to the Canary Islands, Cape Verde, Madeira, and locally in Western Morocco [1,2]. The tree is characterized by a single or multiple trunk growing up to 12 m tall, with a dense umbrella-shaped canopy of thick leaves [3]. It grows slowly, requiring about ten years reaching 1 m tall [4]. Young trees have only a single stem; branching occurs when the tree flowers, when two side shoots at the base of the flower panicle continue the growth as a fork in the stem. Some specimens are believed to be up to 650 years old; the oldest is growing at Icod de los Vinos in Northwest Tenerife [1]. Recently-discovered wild populations in Western Morocco have been described as a separate subspecies, *Dracaena draco* subsp. ajgal, while those on Gran Canaria are sometimes distinguished as a separate species *Dracaena tamaranae* [5]. Rising interest in *Dracaena draco* for its medicinal properties [6] is prompting innovative approaches for efficient use of this endangered tree for food industry and pharmacological applications.

In view of the drastic reduction in the number of individuals and the consequent loss of dragon tree genetic diversity caused by absence of natural regeneration, micropropagation offers many advantages because it potentially can facilitate large-scale production of valuable clones and allow plant reintroduction in its natural ecosystem [7–11]. In addition, micropropagation may be an essential step to obtain plants from frozen collections of dragon tree plant material, the most valuable way of preserving this genus. Unfortunately, most of the published protocols are poorly described, in particular concerning the difficult stage of acclimatization. Therefore, the aim of the present study was to develop an efficient procedure for micropropagation and subsequent acclimatization of Dragon tree (*Dracaena draco*) under in vitro culture conditions. In this study, suitable explant sources, combinations of plant growth regulators for shoot development, and optimal concentrations of IBA for root induction were investigated.

2. Materials and Methods

2.1. Plant Material

Dracaena draco trees purchased from a French Nursery (Annemasse, France) were used as starting material for isolation of dormant buds. They were maintained in greenhouse at 25 ± 2 °C with a 16-h photoperiod. Light intensity was set at 40 μmol m^{-2} s^{-1} and was provided by mercury fluorescent lamps. Additional trees were obtained either by in vitro or in soil germination of fresh *D. draco* seeds and their growth in a greenhouse for 2 years (Figure 1a,b). They were also used as starting material for the isolation of dormant buds. Dormant buds obtained from mature *D. draco* trees (Figure 1c) were treated with 70% ethanol for 5 min and used as starting plant material.

Figure 1. Culture of *D. draco* in soil or in vitro: (**a**) in vitro germination of *D. draco* seeds; (**b**) germination of *D. draco* seeds in soil; (**c**) one-year-old *D. draco* plants obtained from *in vitro* and in soil germinated seeds.

2.2. Dormant Bud Sterilization

Buds were sterilized for varying periods (50 buds per each time of 0.5, 1, 2, 5, 10, and 20 min) in 2.5% HClO (*w/v*) bleach. Three double distilled water washes (for 10 min) were then performed. The washed buds were transferred to Murashige and Skoog (MS) medium (1.5% sucrose, 0.8% agar, pH 5.8; Sigma-Aldrich, Buchs, Switzerland) [12] and bud contamination was recorded after 3 d of culture. Effective viability of the buds was also checked visually by the absence of tissue damage and visible necroses.

Sterile dormant buds were used in the different experiments for testing shoot and root induction media. All inoculations were performed under aseptic conditions in a sterile cabinet. The medium was adjusted to pH 5.8 prior to autoclaving at 121 °C for 25 min. The cultures were maintained at 25 ± 2 °C with a 16-h photoperiod as described earlier. Explant contamination and survival rates were evaluated 5 days after inoculation and percentages were calculated.

2.3. Shoot Induction

Dormant buds recovered from mature *D. draco* trees were used to initiate shoot development on MS medium [12] supplemented with different combinations of kinetin (KIN) or 6-benzylaminopurine (BAP) and naphthaleneacetic acid (NAA) or indole-3-butyric acid (IBA). Media were: S1) 1 mg/L KIN and 1 mg/L NAA; S2) 3 mg/L KIN and 1 mg/L IAA; S3) 1 mg/L BAP and 2 mg/L IBA; and S4) 1 mg/L BAP and 1 mg/L NAA were used [13]. There were 50 buds per growth regulator combination, and all experiments were performed in triplicate. Containers used in the experiment consisted of 30 × 160 mm test tubes. Shoot development data were taken after one month of culture. Developing shoots were maintained and used for rooting experiments.

2.4. Rooting of Developing Shoots

Developing shoots were sub-cultured on MS medium containing different concentrations of IBA; R1) 0 mg/L IBA; R2) 0.5 mg/L IBA; R3) 1 mg/L IBA; and R4) 2 mg/L IBA were used. There were 50 buds per growth regulator, and all experiments were performed in triplicate. Container types used in the experiment consisted of 30 × 160 mm test tubes. Root development data were taken after one month of culture.

2.5. Acclimatization of Rooted D. Draco In Vitro Grown Plants

Acclimatization assays performed in this study used plants rooted in R4 medium followed by culture in MS medium without growth regulators [13]. Single plants were transferred to 500 mL pots, with an autoclaved soil mixture of peat and perlite (3/2, *v/v*), and placed in a glass chamber located in the greenhouse. In the first few days, plants were fogged with sterile water to avoid leaf fade. Time between fogging periods was progressively lengthened to diminish relative humidity. In two independent experiments, plants were transferred to soil and results were examined after 6 weeks.

2.6. Statistical Analysis

The statistical analysis of the data was performed using analysis of variance (ANOVA), using % data from each experiment as a replication. When significant effects ($P < 0.05$) were detected, the treatments were compared using a post-hoc Tukey's honestly significant difference (HSD) test at $P < 0.05$. The statistical program used was IBM SPSS Statistics v. 22. Rooted shoots were maintained and used for acclimatization experiments.

3. Results and Discussion

3.1. Optimization of Dormant Bud Sterilization

The effect of different sterilization periods on bud health and absence of contamination was studied. Table 1 shows that a 2- to 5-min sterilization was optimal for avoiding tissue damage and providing sterile explants (Table 1). Subsequently all sterilization was performed for 3 min. These results are in the same range than those reported by Miller and Murashige [7] for *D. godseffiana*, Vinterhalter [8] for *Dracaena fragrans*, Blanco et al. [11] for *Dracaena deremensis*, and Liu et al. [13] for *Dracaena surculosa*.

Table 1. Effect of sterilization time on infection and survival rates of buds cultured in vitro.

Sterilization Time (min)	Infection (%)	Healthy Buds (%)
20	0 c [z]	6.6 ± 11.5 cd
10	0 c	0 d
5	0 c	32.0 ± 3.6 ab
2	0 c	42.3 ± 2.5 a
1	5.0 ± 3.6 b	21 ± 3.6 bc
0.5	67.2 ± 2.5 a	0 d

[z] Mean separation using Tukey's HSD test at $P \geq 0.05$.

3.2. Shoot Induction

The effect of different treatment combinations of KIN or BAP and NAA or IBA were investigated. Shoot induction results are shown in Table 2. Shoot induction rates varied among the tested media. The highest shoot induction was obtained on media S1 and S2. Shoot induction in the medium S1 was high with up to 80% of the buds developing shoots. Shoot induction on medium S2 was also high approaching a rate of 75% shoot development. Figure 2a–c illustrate an example of developing shoots. Because S1 medium induced the highest shoot induction, this medium was chosen as the standard medium for shoot induction of all dormant buds.

Media S1 and S2 used in this study were successfully used by Blanco et al. [11] and Miller and Murashige [7] to induce differentiation and multiplication of *D. deremensis* and *D. godseffiana*, respectively. Results obtained in this study indicate that these media were also effective in inducing shoot differentiation in *D. draco*.

Table 2. Effect of media hormonal composition on shoot and root induction of *D. draco* buds.

Media	Hormonal Composition	Shooting %	Rooting %
S1	1 mg/L KIN, 1 mg/L NAA	80.6 ± 1.2 a [z]	0 c
S2	3 mg/L KIN, 1 mg/L IAA	76.0 ± 4.6 a	0 c
S3	1 mg/L BAP, 2 mg/L IBA	24.6 ± 1.5 b	0 c
S4	1 mg/L BAP 1 mg/L NAA	0 c	0 c
R1	0 mg/L IBA	0 c	0 c
R2	0.5 mg/L IBA	0 c	0 c
R3	1 mg/L IBA	0 c	51.0 ± 3.6 b
R4	2 mg/L IBA	0 c	66.2 ± 2.0 a

[z] Mean separation using Tukey's HSD test at $P \geq 0.05$.

Figure 2. In vitro multiplication and acclimatization of *D. draco*: (**a**) dormant buds collected from 1-year-old *D. draco* plants in culture; (**b–d**) development of buds after 3, 6 and 9 weeks of culture in S1 medium, respectively; (**e,f**) rooted plants in culture; (**g**) one-year-acclimated plants from *in vitro* micro-propagated *D. draco* plants.

Although having a much lower % shoot induction, medium S3 induced about 24% shoots. This medium has been successfully used by Vinteralter [8] to induce shoot multiplication in *D. fragrans*. Similar results were also obtained by Perez et al. [14] and Aslam et al. [15] using *Dracaena sanderiana*.

3.3. Root Induction and Plant Maintenance

Fifty developing shoots obtained on the medium S1 were cultured on different concentrations of IBA for rooting. After a month of culture, root development occurred abundantly in media R3 and R4 (Table 2). Root formation on high concentration of IBA (1–2 mg/L) had higher rhizogenic potential with harder, longer and whiter roots (Figure 2d,e). The best results for rooting were obtained in the medium containing 2 mg/L IBA. Over 50% and 66% of developing shoots produced roots media R3 and R4, respectively, after a month of culture. It is clear from the data in Table 2 that poor root development occurred with low concentrations of IBA. Better results were obtained in at the higher concentrations of IBA rather than the lower concentrations where no rooting occurred. R4 medium was chosen as the optimum medium for rooting in this study.

These results are in disagreement with those of Vinterhalter [8], who reported that during root induction in *D. fragrans*, 0.5 mg/L IBA was optimal. Parallel with these results, Blanco et al. [11] reported that a hormone-less medium was convenient for rooting *D. deremensis*. These results seem to indicate that rooting in the genus *Dracaena* is variable among species [8,11,13–16].

3.4. Acclimatization of Regenerated Plants

From the plants that rooted in R4 medium (25) and that were transferred to the soil mixture, 100% were totally established after 6 weeks. No morphological differences were found between these plants and the mother plant (Figure 2f). Checking the plants after 18 months for morphological abnormalities revealed no difference with plants obtained from soil grown sites (Figure 2g). In our work, acclimatization was given particular attention. The use of the glass chamber in the greenhouse and fogging during the first days with longer spacing between fogging periods to gradually diminish relative humidity could explain the complete success of acclimatization. Acclimatization success in our study was higher than reported in earlier studies of Vinterhaler [8] for *D. fragrans*, Blanco et al. [11] for *D. deremensis*, and Miller and Murashige [7] for *D. godseffiana*, where low success rates were reported. However, our acclimatization success rates were in the same range reported by Aslam et al. [15] for *D. sanderiana* Sander ex Mast and by Liu et al. [13] for *D. surculosa*. Similar to results of Liu et al. [13] and Aslam et al. [15] using *D. sanderiana* and *D. surculosa*, respectively, ex vitro established plantlets exhibited no visible phenotypic aberrations.

4. Conclusions

In the present study, the highest shoot induction rate, the greatest percent rooting and acclimatization rates were obtained using the S1 and R4 media containing [1 mg/L KIN, 1 mg/L NAA] and [2 mg/L IBA], respectively. Ex vitro transfer of plantlets to soil resulted in no visible phenotypic aberrations as required for efficient multiplication of the tree. The present in vitro culture procedure yielding up to 8 plants per mature plant (data not shown) could be used for micropropagation of *D. draco* with the aim of reintroduction of this endangered species in its native ecosystem where natural regeneration is no longer observed. This in vitro culture procedure offers also the potential for efficient use of *Dracaena draco* by food and pharmacological industries, allowing efficient biotechnological valorization of the species and increasing awareness for its conservation.

Author Contributions: Conceived and designed the experiments: A.G., A.C.B., and L.B. Performed the experiments: A.G., A.C.B., and L.B. Analyzed the data: A.G., L.B., and A.C.B. Contributed reagents/materials/analysis tools: A.G. and L.B. Wrote and enriched the literature: A.G., L.B., and A.C.B.

Funding: This research received no external funding.

Acknowledgments: We are grateful to Maria Dalila Santos (Botanical Gardens of Ajuda, Portugal) for the generous gift of fresh *Dracaena draco* seeds.

Conflicts of Interest: The coauthors declare no conflict of interest.

References

1. Krawczyszyn, J.; Krawczyszyn, T. Photomorphogenesis in *Dracaena draco*. *Trees* **2016**, *30*, 647–664. [CrossRef]
2. Benabid, A.; Cuzin, F. *Dragon tree* (*Dracaena draco* subsp. *ajgal* Benabid et Cuzin) populations in Morocco: Taxonomical biogeographical and phytosociological values. *C. R. Acad. Sci.* **1997**, *320*, 267–277. [CrossRef]
3. Chen, J.; Mc Connell, D.; Henny, R.; Everitt, K. Environmental Horticulture Department, Institute of Food and Agricultural Sciences, University of Florida. 1998. Available online: http://edis.ifas.ufl.edu (accessed on 20 June 2019).
4. Monteiro, A.; Vasconcelos, T.; Tapadabertelli, A. Seed propagation of *Dracaena draco*. *Garcia Orta Sér. Bot.* **1999**, *14*, 187–189.
5. Marrero, A. A new species of the wild dragon tree, *Dracaena* (Dracaenaceae) from Gran Canaria and its taxonomic and biogeographic implications. *Bot. J. Linn. Soc.* **1998**, *128*, 291–314.
6. Jura-Morawiec, J.; Tulik, M. Dragon's blood secretion and its ecological significance. *Chemoecology* **2016**, *26*, 101–105. [CrossRef]
7. Miller, L.R.; Murashige, T. Tissue culture propagation of tropical foliage plants. *In Vitro Cell. Dev. Biol. Anim.* **1976**, *12*, 797–813. [CrossRef]
8. Vinterhalter, D. In vitro Propagation of Green-Foliaged Dracaena-Fragrans Ker. *Plant Cell Tissue Org.* **1989**, *17*, 13–19.
9. Vinterhalter, D.; Vinterhalter, B. Micropropagation of *Dracaena* Species. In *Biotechnology in Agriculture and Forestry*; Springer: Berlin/Heidelberg, Germay, 1997; Volume 40, pp. 131–146.
10. Tian, L.; Tan, H.Y.; Zhang, L. Stem segment culture and tube propagation of *Dracaena saneriana* cv. virscens. *Acta Hortic. Sin.* **1999**, *26*, 133–134.
11. Blanco, M.; Valverde, R.; Gomez, L. Micropropagation of *Dracaena deremensis*. *Agron. Costarric.* **2004**, *28*, 7–15.
12. Murashige, T.; Skoog, F. A Revised Medium for Rapid Growth and Bio Assays with Tobacco Tissue Cultures. *Physiol. Plant.* **1962**, *15*, 473–497. [CrossRef]
13. Liu, J.; Deng, M.; Henny, R.J.; Chen, J.; Xie, J. Regeneration of *Dracaena surculosa* through indirect shoot organogenesis. *HortScience* **2010**, *45*, 1250–1254. [CrossRef]
14. Pérez, G.H.Y.; Claret, M.C.C.; Ada, M.M.M. In vitro morphogenesis of Dracena. *Agron. Trop.* **2006**, *56*, 577–583.
15. Aslam, J.; Mujib, A.; Sharma, M.P. In vitro micropropagation of *Dracaena sanderiana* Sander ex Mast: An important indoor ornamental plant. *Saudi J. Biol. Sci.* **2013**, *20*, 63–68. [CrossRef]
16. Junaid, A.; Mujib, A.; Sharma, M.P. Effect of growth regulators and ethylmethane sulphonate on growth, and chlorophyll, sugar and proline contents in *Dracaena sanderiana* cultured in vitro. *Biol. Plant.* **2008**, *52*, 569–572. [CrossRef]

 *horticulturae*

Article

Improved Propagation and Growing Techniques for Oleander Nursery Production

Leo Sabatino, Fabio D'Anna and Giovanni Iapichino *

Dipartimento Scienze Agrarie, Alimentari e Forestali, Università di Palermo, 90128 Palermo, Italy
* Correspondence: giovanni.iapichino@unipa.it; Tel.: +39-091-23862215

Received: 30 June 2019; Accepted: 30 July 2019; Published: 1 August 2019

Abstract: In the first trial, we examined rooting of stem cuttings in relation to number of nodes and indole-3-butyric acid (IBA) treatment in several *Nerium oleander* clones grown in Sicily. In a second trial, we tested the effect of different forcing dates and shading on oleander plants for gardens and natural landscapes. Three- and four-node cuttings, ranging in length from 10 to 14 cm, were significantly superior to two-node cuttings (8–10 cm long) in terms of rooting percentage and number of roots per cutting. The application of IBA improved rooting percentage and root number as compared to untreated control. Irrespective of IBA, rooting percentages ranged from 94% in clone 1 to 52% in clone 4. Shaded plants forced in October were significantly higher than those forced in November and in December. Beginning of flowering was delayed in unforced plants. Plants forced in October flowered significantly sooner (first decade of March) than unforced ones (first decade of May) and reached complete flowering almost two months earlier (last week of March).Shading had little effect on plants forced in October and in November as compared to unshaded plants in terms of start of flowering, but it slightly hastened beginning of flowering of December forced plants as compared to their unshaded counterparts.

Keywords: rooting; cutting; forcing; oleander; shading

---

1. Introduction

Nerium oleander L. (oleander) is a shrub native to northern Africa and the Mediterranean region grown for flowers and evergreen foliage. The long flowering period extending from early spring through late fall and the appealing flower display make oleander a valuable ornamental plant and one of the best shrubs for landscaping and xero-gardening projects in semi-arid environments [1]. The species has also become an important plant for flowering pots [2,3].

Implementing propagation methods to enhance transplant success, establishment, and post-plant maintenance is a major objective for plant nurseries involved in the production of shrubs to be used for gardens and natural landscapes in regions with a Mediterranean climate. In this regard, the production of oleander rooted cuttings with a well-developed root system is fundamental for successful transplanting and establishment in the field. By growing plants in a greenhouse or other protected facilities, plant nurseries force plants to bloom out of season. Ornamental shrubs with colors create a great amount of impulse buying from the average garden plant consumer because a flowering plant provides immediate gratification and guarantees that the consumer is getting what they paid for [4]. Based on these considerations, reducing the time span between propagation and flowering by individuating opportune forcing and shading techniques induces oleander to bloom out of season and improves its marketability for garden plantings.

According to Hartmann [5], oleander cultivars are clonally propagated by leafy cuttings, which root easily if taken from rather mature wood during the summer and treated with a 3000 ppm indole-3-butyric acid (IBA) quick-dip. Toogood [6] affirms that to produce a oleander flower plant in 2 years, leafy cuttings or semi-hardwood cuttings can be rooted directly in 3–6 weeks and that bottom-heat at 12–20 °C enhances rooting. Dirr [7] suggests to root cuttings collected from mature wood in late July or August after a 3000 ppm IBA application. Ochoa et al. [8,9] report that basal temperatures ≥25 °C favor rooting and root quality of semi-hardwood oleander cuttings, and that cuttings taken from the basal part of the stem produce larger root growth, although more roots with a more homogeneous length distribution were obtained from distal cuttings. However, to our knowledge, no published data is available concerning the influence of the morphological characteristics of the cuttings on adventitious root formation. Therefore, in the first series of experiments our objective was to examine rooting of stem cuttings in relation to number of nodes and IBA treatment in several *Nerium oleander* clones grown in Sicily.

Although there have been several studies undertaken to improve the suitability of this species as a potted plant, especially through the use of plant growth regulators, there are no reports concerning the effects of shading and diverse forcing date on oleander nursery production. Therefore, in the second experiment we tested the effect of different forcing dates and shading on oleander production.

2. Materials and Methods

The research was conducted at the experimental field of the Department of Agricultural, Food and Forest Sciences of Palermo (SAAF), at Marsala, Trapani Province (longitude 12°26′ E, latitude 37°47′ N, altitude 37 m above sea level (asl) in the north-western coast of Sicily (Italy).

2.1. Propagation Treatments

In the first experiment, distal 30-cm-long stem cuttings were harvested on 23 March 2018, from actively growing oleander stock plants. Plants were of cutting origin and from a 10-year-old Sicilian unnamed clone characterized by a red corolla (clone 1). Hardwood stem cuttings were collected from the previous year's growth cycle in the middle of the crown of the stock plants. Cuttings were stored overnight at 6 °C. The next day, the terminal shoot was removed and cuttings were cleaned and defoliated prior to planting. The cuttings were trimmed into three types: (1) two-node cuttings (length 8–10 cm); (2) three-node cuttings (length 10–12 cm); (3) four-node cuttings (length 12–14 cm). Average diameter of the cuttings ranged from 10 to 13 mm. All cuttings had the first cut made 10 mm above the distal node and the second cut made 20 mm below the proximal node. To verify the cutting response to exogenous auxin, cuttings were dipped to a 2.0 cm depth in a 0.3% indole-3-butyric acid (IBA) solution for 30 s. Reagent grade IBA (Sigma Aldrich, Italy) was dissolved in 90% ethanol, then brought to a final concentration of 6% ethanol and 94% distilled water (*v/v*). Untreated cuttings were dipped in distilled water (control). After 30 and 45 d, data were recorded as percent survival, percentage of cuttings rooted, number of roots per cutting, and length of the 6 longest roots. Growth resulting from sprouting of the buds present on the original cutting material was rated by measuring shoot length after 30 and 60 d.

In the second experiment, distal 30-cm-long hardwood stem cuttings were harvested on 30 March 2018, from actively grown stock plants of five Sicilian unnamed oleander clones characterized by a red (clone 1), pink (clone 2), white (clone 3), salmon (clone 4), and yellow (clone 5)corolla (Figure 1).

Cuttings were collected as described in the first experiment, and stored overnight at 6 °C. The next day, three-node cuttings (length 10–12 cm) were prepared as described in the first experiment. To verify the cutting response to exogenous auxin, half of the cuttings were dipped to a 2.0 cm depth in a 0.3% IBA solution for 30 s, whereas the remainder were dipped in distilled water (control). After 30 and 45 days, data were recorded as percentage of cuttings rooted, number of roots per cutting, and length of the 6 longest roots. Growth resulting from sprouting of the buds present on the original cutting material was rated by recording shoot length and number of leaves per rooted cutting after 30 and 60 d.

Figure 1. *Nerium oleander* clones tested.

In the two experiments, propagation was performed in an unheated greenhouse covered with clear polyethylene (PE) and external 70% shade-cloth. Single cuttings were stuck to each plastic pot (diameter 12) containing a peat:perlite (2:1/*v:v*) media. Air temperature in the greenhouse was 18–23 °C during the day and 14–16 during the night. Rooting medium temperature was 18–22 °C during the day and 14–18 °C during the night. Intermittent mist operated daily for 30 s every 2 h from 8:30 A.M. to 6:00 P.M.

2.2. Forcing and Shading Treatments for Plant Nursery Production

In this experiment, three-year-old plants of a Sicilian unnamed clone characterized by a red corolla (Clone 1) were used. These plants were originally from cuttings and were grown in round plastic pots (20 cm diameter) filled with a commercial peat-perlite substrate mix (VIGORPLANT, SER FS V10-18, Italy) 80:20 (*v/v*). The substrate mix used had the following properties: pH 5.5, electric conductivity 0.20 dS^{m-1}, (dry) bulk density 95 kg m^{-3}, total porosity 94%, NH$_4$ 220 mg kg^{-1}, NO$_3$ 833 mg kg^{-1}, P 40 mg kg^{-1}, K 631 mg kg^{-1}. In March 2018, to obtain homogenous plants the material plants were pruned to five branches. Seven days before the beginning of the experiment, plants were fertilized with a water soluble fertilizer (20-20-20) at 2 gL^{-1}. In the first week of June 2018, to verify the plant response to shading, half of the plants were placed into a shade house (Aluminet TM 30% shade cloth), whereas the remaining were left outside. To test the effect of different forcing times, three forcing dates were selected: 1 October, 1 November, and1 December. For each shading treatment (shading vs.no shading), half of the plants were transferred into ethylene vinyl acetate (EVA)-covered forcing tunnels and half of the plants were left outside. The experimental plants were watered as necessary to ensure that the plants would not be exposed to drought stress. Maximum, minimum, and average temperature were monitored during the forcing periods (Figure 2).

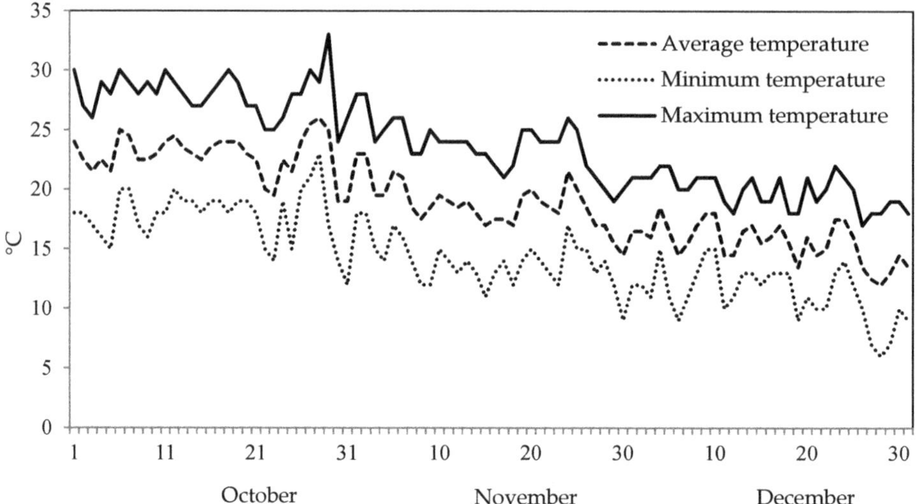

Figure 2. Maximum, minimum, and average temperatures during the forcing periods.

In order to assess the ornamental quality of oleander plants, data were collected on plant height at beginning of flowering, number of days from first of January to beginning of flowering, and number of days from first of January to full flowering.

2.3. Statistical Analysis

In the first propagation experiment, three levels of node number (two-, three-, and four-node cuttings) in combination with IBA concentrations of 0 and 0.3% were tested in a two-factor, randomized complete block design with 4 blocks and 96 cuttings per IBA treatment, 24 per block. Data were subjected to a two-way analysis of variance ANOVA. In the second propagation experiment a 5 × 2 (clones × IBA) factorial set of treatments within a complete randomized block design was used with 4 blocks and 96 cuttings per IBA treatment. Data were subjected to a two-way ANOVA. In the forcing experiment, 2 levels of shading (shading vs. unshaded), in combination with 3 forcing times plus the not-forced treatment were tested in a 2 × 4 factorial set of treatments within a randomized complete block design with 4 blocks and 100 plants per treatment. Data were subjected to a two-way ANOVA.

All data were statistically analyzed using the Statistical Package for Social Science (SPSS) software version 14.0 (StatSoft, Inc., Chicago, IL, USA). Mean separation was assessed by Tukey Honest Significance Difference (HSD) test. Percentage data were subjected to arcsin transformation before ANOVA analysis ($\varnothing$ = arcsin $(p/100)^{1/2}$).

3. Results

3.1. Propagation Treatments

In the first experiment, overall percent survival was 97% to 98%; no significant effects of node number and IBA treatment were found on cutting survival (data not shown). Rooting percentage at 30 days after cutting insertion into the rooting medium was significantly affected by node number and IBA treatment (Table 1). Regardless of auxin treatment, three- and four-node cuttings gave a significantly higher rooting percentage as compared to two-node cuttings. Independently of node number, percentage of rooting increased from 83% in untreated cuttings to 93% in cuttings exposed to IBA. However, no significant interaction was found between node number and IBA exposure. Node number did not affect the number of roots per cutting 30 days after planting regardless of IBA treatment (Table 1), whereas root number per cutting at 45 days averaged over IBA was significantly higher in three and four node cuttings as compared to two node cuttings. Irrespective of node level, IBA significantly increased the number of roots per cutting from 7 to 10 and from 16 to 23 after 30 and 45 days, respectively. No significant interaction was found between node number and IBA exposure in terms of root number.

Table 1. Effects of number of nodes per cutting and indole-3-butyric acid (IBA) on rooting and shoot length of *Nerium oleander*.

Treatments	Rooting (%)	No. of Roots Cutting^{-1}				Root Length (cm)		Shoot Length (cm)		
		at 30 d		at 45 d		at 30 d	at 45 d	at 30 d		at 60 d
Node Number (N)										
2	87.7 b^{z}	7.8		15.9	b	11.8	12.0	6.8		8.5
3	94.1 a	8.8		22.3	a	10.4	12.2	7.0		8.9
4	91.9 a	8.9		21.1	a	12.2	14.0	8.2		8.6
IBA										
0 ppm	83.5 b	6.9	b	16.4	b	10.4	13.1	5.9	b	8.4
3000 ppm	93.0 a	10.0	a	23.2	a	12.5	12.4	8.1	a	8.8
Significance										
N	**^{y}	NS		**		NS	NS	NS		NS
IBA	*	**		***		NS	NS	***		NS
N × IBA	NS	NS		NS		NS	NS	NS		NS

^{z} Data within a column followed by the same letter are not significantly different at $p \leq 0.05$ according to Tukey HSD Test. ^{y} The statistical significance is designated by asterisks as follows: *, at $p \leq 0.05$; **; at $p \leq 0.01$; ***, at $p \leq 0.001$; NS = not significant.

Node number and IBA treatment did not significantly affect root length eitherat 30 or45 d from planting in the rooting medium (Table 1). The length of the shoots resulting from sprouting of the buds present on the original cutting material averaged over IBA was not influenced by node number after either 30 or60 days from planting (Table 1). After 30 days, shoots developed from cuttings exposed to IBA were significantly longer than those grown from untreated ones regardless of the node number. There was no significant interaction between node number and IBA exposure in terms of root length.

In the second experiment, there were significant effects of auxin treatment, tested clones, and their interaction for rooting and shoot growth of the cuttings (Table 2).

Regardless of IBA treatment, clone 1 (red) had the highest rooting percentage (94%), followed by clone 2 (pink), clone 3 (white), and clone 5, whereas, clone 4 (salmon) showed the lowest rooting (52%). Rooting percentage averaged over clones accounted for 78% and 80% in untreated and IBA treated cuttings, respectively. The ANOVA revealed a significant effect of the interaction between clones and IBA (Tables 2 and 3). Rooting percentages ranged from 96% in clone 1 (red color) cuttings exposed to IBA to 48% and 56% in clone 4 cuttings exposed and not exposed to IBA, respectively.

Table 2. Effects of indole-3-butyric acid (IBA) on rooting and growing of three-node cuttings of several clones of *Nerium oleander*.

Treatments	Rooting (%)	No. of Roots Cutting^{-1}		Root Length (cm)		Shoot Length (cm)		No. of Leaves Cutting^{-1}	
		at 30 d	at 45 d	at 30 d	at 45 d	at 30 d	at 60 d	at 30 d	at 60 d
Clones									
Clone 1	93.8 a[z]	8.8 ab	22.3 a	10.4 b	12.1	7.0	8.8 b	16.3 b	28.9 a
Clone 2	90.1 b	10.2 a	14.2 bc	14.7 a	13.5	8.2	10.4 a	19.0 ab	22.4 b
Clone 3	87.5 b	10.2 a	16.5 b	11.6 ab	12.4	9.5	11.0 a	22.1 a	23.0 b
Clone 4	52.1 d	6.3 b	7.3 d	8.4 b	13.3	7.4	8.8 b	15.8 b	23.7 b
Clone 5	71.8 c	7.4 ab	9.3 cd	11.4 ab	10.8	8.9	9.8 ab	19.0 ab	22.9 b
IBA									
0 ppm	78.3	7.3 b	12.9 b	11.8	12.3	7.6	9.4 b	19.5	26.3 a
3000 ppm	79.8	9.9 a	15.3 a	10.8	12.5	8.8	10 a	17.4	22.0 b
Significance									
Clones	***[y]	*	***	**	NS	NS	***	**	***
IBA	NS	**	*	NS	NS	NS	**	NS	***
Clones × IBA	**	NS	NS	NS	NS	NS	NS	*	**

[z] Data within a column followed by the same letter are not significantly different at $p \leq 0.05$ according to Tukey HSD Test. [y] The statistical significance is designated by asterisks as follows: *, at $p \leq 0.05$; **, at $p \leq 0.01$; ***, at $p \leq 0.001$; NS = not significant.

Table 3. Effects of the interaction clones and IBA on rooting and growing of three-node cuttings of several clones of *Nerium oleander*.

Treatments	Rooting (%)		No. of Leaves Cutting^{-1} at 30 Days		No. of Leaves Cutting^{-1} at 60 Days	
Clone 1 × 0 ppm	87.5	c [z]	14.5	c	28.6	a
Clone 1 × 3000 ppm	95.8	a	18.0	ab	29.2	a
Clone 2 × 0 ppm	91.7	b	21.8	ab	24.9	b
Clone 2 × 3000 ppm	92.7	b	16.3	b	19.9	c
Clone 3 × 0 ppm	85.4	c	23.5	a	24.2	b
Clone 3 × 3000 ppm	89.6	bc	20.8	ab	21.8	bc
Clone 4 × 0 ppm	56.3	e	19.0	ab	27.6	a
Clone 4 × 3000 ppm	47.9	f	12.5	c	19.8	c
Clone 5 × 0 ppm	70.8	d	18.5	b	26.3	b
Clone 5 × 3000 ppm	72.9	d	19.5	b	19.5	c

[z] Data within a column followed by the same letter are not significantly different at $p \leq 0.05$ according to Tukey HSD Test.

Irrespective of the auxin treatment, 30 days after cutting insertion in the rooting medium, clone 2 (pink) and clone 3 (white) gave the highest number of roots per cutting (10.0 roots), followed by clone 1 (8.8 roots), clone 5 (7.4 roots), and clone 4 (6.3 roots) (Table 2). Data expressed as number of roots per cutting 45 days after planting, regardless of IBA, revealed that clone 1 had the highest score because it almost tripled root number. Data from the remaining treatments supported the trend established at 30 days.

Irrespective of the clones, IBA exposure significantly increased root number as compared to the control, both at 30 and 45 days after planting. There was no significant interaction between node number and IBA exposure in terms of root number, either at 30 or 45 days after planting.

Data on root length at 30 d averaged over IBA showed significant effects of clone only. Clone 2 had the longest roots, followed by clone 3 and clone 5. Irrespective of the clones, IBA treatment had no effect on root length. No significant interaction was found between IBA and clones in terms of root length. There were no significant effects of auxin treatment, tested clones, and their interaction for root length at 45 days.

Treatments had no significant effects on shoot length after 30 days from planting (Table 2). However, after 60 days from planting, irrespective of IBA, shoot length was significantly affected by clones. Clone 2 (pink) and clone 3 (white) had the longest shoots, whereas clone 1 (red) and clone 4 (salmon) had the shortest ones (Table 2). Shoot length averaged over clones was higher in IBA treated cuttings as compared to the untreated ones.

Irrespective of the auxin treatment, clones significantly influenced number of leaves per cutting^{-1} at 30 days from planting (Table 2). Clone 3 (white) had the highest number of leaves and clones 1 and 4 the lowest. On the contrary, regardless of the clones, IBA did not significantly affect number of leaves at 30 days from planting. A significant interaction was found between clones and IBA in terms of number of leaves per cutting after 30 days (Table 3); the highest number of leaves per cutting was observed in clone 3 and clone 2 cuttings in absence of IBA, followed by clone 3 cuttings exposed to IBA.

There were highly significant effects of tested clones, auxin treatment, and their interaction for number of leaves per cutting after 60 days from planting (Table 2). Regardless of IBA, clone 1 had the highest number of leaves (29 leaves), followed by clone 4 (24 leaves). Number of leaves per cutting averaged over clones was higher in IBA-treated cuttings as compared to the untreated ones.

Regarding the interaction, the highest number of leaves per cutting was observed in clone 1, either in the presence or in absence of IBA treatment, whereas the lowest value was detected in the combination of clone 5 with IBA (Table 3).

3.2. Forcing and Shading Treatments

There were significant effects of forcing dates, shading, and their interaction for plant height (Table 4).

Table 4. Effects of the interaction forcing and shading on plant height, beginning of flowering, and full flowering of *Nerium oleander* pot plants.

Treatments	Plant Height (cm)		Beginning of Flowering (Days from 1 January)		Full Flowering (Days from 1 January)	
Not-forced × U	82.3	d [z]	122.7	a	142.0	a
October forcing × U	73.0	e	69.0	e	85.0	g
November forcing × U	95.0	c	78.3	d	112.7	e
December forcing × U	97.7	c	111.7	b	130.0	c
Not-forced × S	98.0	c	120.0	a	136.0	b
October forcing × S	116.3	a	71.0	e	96.7	f
November forcing × S	103.0	b	79.3	d	95.0	f
December forcing × S	105.0	b	99.0	c	117.0	d
			Significance			
Forcing (F)	***[y]		***		***	
Shading (S)	**		*		NS	
F × S	***		*		***	

[z] Data within a column followed by the same letter are not significantly different at $p \leq 0.05$ according to Tukey HSD Test. [y] The statistical significance is designated by asterisks as follows: *, at $p \leq 0.05$; **; at $p \leq 0.01$; ***, at $p \leq 0.001$; NS = not significant.

Plants exposed to shading from June to September and transferred into the tunnels on October 1 were significantly higher than shaded plants forced on November 1 and December 1. These plants in turn were significantly higher than their unshaded counterparts. Plants in all other treatments were significantly shorter (73 to 98 cm in height).

Regardless of shading, forcing hastened beginning of flowering, as plants transferred into the tunnels in October, November, and December flowered significantly earlier than unforced ones. Regarding the interaction, both shaded and unshaded plants forced in October began to flower 71 and 70 days after January 1, respectively, whereas shaded and unshaded plants forced in November began to flower 9 and 8 days later, respectively. Plants in the remaining plots began to flower significantly later.

Irrespective of shading, forced plants reached full flowering significantly earlier than unforced ones. There was a significant interaction between forcing and shading in terms of full flowering. Unshaded plants forced in October reached the full flowering phase 11 and 10 days earlier than shaded plants transferred into the tunnels in October and in November, respectively. Plants in the remaining treatments reached full flowering significantly later.

4. Discussion

In the first experiment of the present study, three- and four-node cuttings, ranging in length from 10 to 14 cm, were significantly superior to two-node cuttings (8–10 cm long) in rooting percentage and number of roots per cutting. Our results agree with those obtained by Smalley and Dirr [10], Henry et al. [11], Hinesley et al. [12], and Caruso and Iapichino [13], who reported that longer and multi-node cuttings positively affect root count in *Acer rubrum*, *Juniperus virginiana*, *Chamaecyparis thyoides*, and *Plumeriarubra*, respectively, as compared to shorter cuttings. The superior efficacy of longer cuttings in inducing higher rooting performance as compared to shorter cuttings might be attributed, as suggested by Beyl et al. [14] in a study on *Actinidia arguta*, to their higher carbohydrate reserves.

In our study, the application of IBA improved rooting percentage and root number as compared to untreated control. IBA has been reported to increase in vivo and in vitro adventitious root formation in many species, including vegetables, perennials, and ornamental shrubs [15–26].

In the second experiment in the absence of IBA, clones 1, 2, and 3 had a rooting percentage ranging from 85% to 92%, whereas in clones 4 and 5, rooting did not exceed 70%. These findings partly agree with those obtained by Ochoa et al. [8,9], who reported that in the absence of auxin treatments rooting percentage of semi-hardwood oleander cuttings can be higher than 80%. We also found that exposing clones to IBA resulted in an increase in rooting capacity for clones 1, 2, and 3, but had no effect on clones 4 and 5, whose rooting capacity either declined or remained unchanged. This different response among clones confirms that rooting ability of cutting in oleander is influenced by genotype, as reported for other woody species, such as (*Corylusavellana*) and olive (*Olea europaea*) [25,26].

Shaded plants forced in October were significantly higher than those forced in November and in December, which in turn where higher than plants of the other treatments.

Oleander plants forced in October flowered significantly sooner (first decade of March) than unforced ones (first decade of May) and reached complete flowering almost two months earlier (last week of March). Beginning of flowering was delayed in unforced plants. Shading had little effect on plant forced in October and in November as compared to unshaded plants, in terms of beginning of flower, but it slightly hastened beginning of flowering in December forced plants as compared to their unshaded counterparts.

5. Conclusions

According to our findings we suggest that the use of three- and four-node cuttings together with IBA basal dip may be beneficial to propagators wishing to produce oleander with a well-developed root system and capable of achieving rapid establishment in the field. The greater root systems of three- and four-node IBA-treated rooted cuttings as compared to untreated ones facilitated the uptake of water and nutrients and provided more carbohydrate reserves to support early spring sprouting and faster top growth in the subsequent growing season. Shading (from June to September) prior to forcing promoted vigorous growth and taller plants as compared to no shading. Shading might also save a considerable amount of water during the driest months, especially in Mediterranean areas where water is a scarce resource. Finally, by forcing oleander to flower earlier, growers can supply blooming plants to the market when they would naturally still be vegetative. Overall, this would provide nurseries involved in ornamental plant production with a great marketing tool. Therefore we can conclude that the use of appropriate plant propagation material and techniques together with opportune shading and forcing might be beneficial to Sicilian plant nurseries involved in oleander production.

Author Contributions: L.S., F.D., and G.I. conceived and designed the research. L.S. carried out field work and analyzed the data. G.I. and L.S. wrote the manuscript. All authors read and approved the manuscript.

Funding: This research received no external funding.

Conflicts of Interest: The authors declare no conflict of interest.

References

1. Franco, J.A.; Martínez-Sánchez, J.J.; Fernández, J.A.; Bañón, S. Selection and nursery production of ornamental plants for landscaping and xerogardening in semi-arid environments. *J. Hortic. Sci. Biotechnol.* **2006**, *81*, 3–17. [CrossRef]
2. Eggenberger, R.; Eggenberger, M.E. *The Handbook of Oleanders*; Tropical Plant Specialists: Cleveland, GA, USA, 1996.
3. Lenzi, A.; Pittas, L.; Martnelli, T.; Lombardi, P.; Tesi, R. Response to water stress of some oleander cultivars suitable forr pot plant production. *Sci. Hortic.* **2009**, *122*, 426–431. [CrossRef]
4. Pilon, P. *Perennial Solutions: A Grower's Guide to Perennial Production*, 1st ed.; Ball Publishing: Batavia, IL, USA, 2005; p. 546.
5. Hartmann, H.D.; Kester, D.E.; Davies, F.J.; Geneve, R.L. *Plant Propagation: Principles and Practices*; Prentice Hall: Upper Saddle River, NJ, USA, 2003; p. 772.
6. Toogood, A.R. *Plant Propagation*; DK Publishing Inc.: New York, NY, USA, 1999.

7. Dirr, M.A.; Heuser, C.W. *The Reference Manual of Woody Plant Propagation*; Varsity Press: Athens, Greece, 2006.
8. Ochoa, J.; Bañón, S.; Fernández, J.A.; Gonzalez, A.; Franco, J.A. Influence of cutting position and rooting media on rhizogenesis in oleander Cuttings. *Acta Hort.* **2003**, *608*, 101–106. [CrossRef]
9. Ochoa, J.; Bañón, S.; Fernández, J.A.; Franco, J.A.; Martínez-Sánchez, J.J. Rooting Medium Temperature and Carbohydrates Affected Oleander Rooting. *Acta Hort.* **2004**, *659*, 239–244. [CrossRef]
10. Smalley, T.J.; Dirr, M.A. Effect of cutting size on rooting and subsequent growth of *Acer rubrum* 'Red Sunset' cuttings. *J. Environ. Hortic.* **1987**, *5*, 122–124.
11. Henry, F.A.; Blazich, P.H.; Hinesley, L.E. Vegetative propagation of Eastern redcedar by stem cuttings. *HortScience* **1992**, *27*, 1272–1274. [CrossRef]
12. Hinesley, L.E.; Blazich, F.A.; Snelling, L.K. Propagation of Atlantic white cedar by stem cuttings. *HortScience* **1994**, *29*, 217–219. [CrossRef]
13. Caruso, S.; Iapichino, G. Basal heat improves adventitious root quality in Plumeria (*Plumeriarubra* L.) stem cuttings of different sizes. *J. Appl. Hortic.* **2014**, *17*, 22–25.
14. Beyl, C.A.; Ghale, G.; Zhang, L. Characteristics of hardwood cuttings influence rooting of *Actinidia arguta* (Siebold and Zucc.) Planch. *HortScience* **1995**, *30*, 973–976. [CrossRef]
15. Blazich, F.A. Chemicals and formulations used to promote adventitious rooting. In *Adventitious Root Formation in Cuttings*; Davis, T.D., Haissig, B.E., Sankhla, N., Eds.; Dioscorides Press: Portland, OR, USA, 1988; pp. 132–149.
16. Iapichino, G. Micropropagation of Globe Artichoke (*Cynara cardunculus* L. var. *scolymus*). In *Protocols for Micropropagation of Selected Economically-Important Horticultural Plants, Methods in Molecular Biology*; Lambardi, M., Ozudogru, E.A., Jain, S.M., Eds.; Springer: New York, NY, USA, 2013; Volume 11013, pp. 329–339.
17. Iapichino, G.; Arnone, C.; Bertolino, M.; AmicoRoxas, U. Propagation of three Thymus by stem cuttings. *Acta Hort.* **2006**, *723*, 411–413. [CrossRef]
18. Kieffer, M.; Fuller, M.P. In vitro propagation of cauliflower using curd microexplants. In *Protocols for Micropropagation of Selected Economically-Important Horticultural Plants, Methods in Molecular Biology*; Lambardi, M., Ozudogru, E.A., Jain, S.M., Eds.; Springer: New York, NY, USA, 2013; Volume 11013, pp. 369–380.
19. Carpenter, W.J.; Cornell, J.A. Auxin application duration and concentration govern rooting of Hibiscus stem cuttings. *J. Am. Soc. Hort. Sci.* **1992**, *117*, 68–74. [CrossRef]
20. Sabatino, L.; D'anna, F.; Iapichino, G. Cutting type and IBA treatment duration affect *Teucrium fruticans* adventitious root quality. *Not. Bot. Horti Agrobot. Cluj Napoca* **2014**, *42*, 478–481. [CrossRef]
21. Kashefi, M.; Zarei, H.; Bahadori, F. The Effect of Indole Butyric Acid and the Time of Stem Cutting Preparation on Propagation of Damask Rose Ornamental Shrub. *JOP* **2014**, *4*, 237–243.
22. Kaviani, B.; Gholami, S. Improvement of Rooting in Forsythia×intermedia Cuttings by Plant Growth Regulators. *JOP* **2016**, *6*, 125–131.
23. Pacholczak, A. The effect of the auxin application methods on rooting of *Physocarpus opulifolius* Maxim. cuttings. *Propag. Ornam. Plants* **2015**, *15*, 147–153.
24. Zhang, L.; Wang, S.; Guo, W.; Zhang, Y.; Shan, W.; Wang, K. Effect of Indole-3-Butyric Acid and rooting substrates on rooting response of hardwood cuttings of *Rhododendron fortune* Lindl. *Propag. Ornam. Plants* **2015**, *15*, 79–86.
25. Cristofori, V.; Rouphael, Y.; Rugini, E. Collection time, cutting age, IBA and putrescine effects on root formation in *Corylus avellana* L. cuttings. *Sci. Hortic.* **2010**, *124*, 189–194. [CrossRef]
26. Sebastiani, L.; Tognetti, R. Growing season and hydrogen peroxide effects on root induction and development in *Olea europaea* L. (cvs 'frantoio' and 'gentile di larino') cuttings. *Sci. Hortic.* **2004**, *100*, 75–82. [CrossRef]

 *horticulturae*

Article

LEDs Combined with CHO Sources and CCC Priming PLB Regeneration of *Phalaenopsis*

Hasan Mehraj [1,2,*], **Md. Meskatul Alam** [1,2], **Sultana Umma Habiba** [1,2] **and Hasan Mehbub** [2]

1   The United Graduate School of Agricultural Sciences, Ehime University, Ehime 790-8577, Japan;
     alam_ju87@yahoo.com (M.M.A.); bably.sau04@gmail.com (S.U.H.)
2   Faculty of Agriculture, Kochi University, Kochi 783-8502, Japan; sujanhasan@gmail.com
*   Correspondence: hmehraj02@yahoo.com

Received: 19 February 2019; Accepted: 17 April 2019; Published: 6 May 2019

Abstract: Throughout this study, the objective was to determine the most effective carbohydrate (CHO) sources under different light-emitting diodes (LEDs), and the impact of chlorocholine chloride (CCC), for the in vitro regeneration of the protocorm-like bodies (PLBs) in *Phalaenopsis* 'Fmk02010'. We applied 15 LEDs combined with three CHO sources and five CCC concentrations in the study. Organogenesis of PLBs was very poor in maltose both for the number of PLBs and their fresh weight (FW) compared to media containing sucrose and trehalose. Sucrose was the best CHO source under the red-white (RW) LED for the in vitro organogenesis of PLBs (PLBs: 54.13; FW: 0.109 g), while trehalose was best under the blue-white (BW) LED (PLBs: 36.33, FW: 0.129 g). The red-blue-white (RBW)-trehalose combination generated a suitable number of PLBs (35.13) with the highest FW (0.167 g). CCC at 0.01, 0.1, and 1 mgL^{-1}CCC had no effect on PLB formation or FW, but 10 mg L^{-1} reduced both. RW-sucrose, BW-trehalose, and RBW-trehalose were the best combinations for PLB organogenesis. The addition of low concentrations of CCC in the plant culture medium are unnecessary.

Keywords: protocorm-like bodies; light-emitting diode; trehalose; maltose; CCC; correlation; growth retardants

---

1. Introduction

Phalaenopsis is the most important and valuable commercial orchid in the Orchidaceae family. It is widely accepted both as cut and pot flowers. Unlike most flowering plants, orchids have a very unique reproductive system. Propagation of *Phalaenopsis*, either vegetatively or by seed, is quite difficult. Tissue culture is the common method due to its successful and rapid propagation. PLB regeneration is the best and most efficient technique for orchid micropropagation [1], because it has a rapid proliferation capacity for producing a large number of protocorm-like bodies PLBs within a short period [2]. They can be induced directly from explants, such as shoot tips [3], flower stalk buds [4], root tips [5], and leaf segments [6]. The indirect regeneration of PLBs can be done by embryogenic callus culture using solid [7] or liquid [8] suspension cultures. Proper media compositions with optimum culture conditions are among the significant factors for fast and high quality plantlet regeneration through PLBs [9,10]. PLBs are the sole form of somatic embryo that imitates the zygotic embryo in natural seed, but unlike the zygotic embryo, it can grow continuously without any dormancy [11].

Media ingredients are the key factor for successful PLB regeneration in vitro. Plant tissue culture media generally have mineral salts, vitamins, growth regulators, and water [12]; another important component is the carbon source to supply energy [13]. There are many carbon sources like sucrose, fructose, glucose, trehalose, maltose, and sorbitol [8,14] used for plant tissue culture that might be in simple or complex forms [15]. It is well known that plants are sensitive to light. Light also affects PLB regeneration through photosynthetic and phototropic responses and may depend on light quality and photoperiod [16]. Fluorescent lights are commonly used. LED lights are currently used for in vitro cultivation. Power consumption of fluorescent lights is greater, and they produce a wide range of wavelengths (350–750 nm) that are unnecessary for plant development. Monochromatic LEDs emit light at specific wavelengths. LEDs are used commercially in plant tissue culture due to their numerous advantages compared to conventional light systems; such as wavelength specificity, durability, small size, long operating time, relatively cool emitting surface, and the ability to control spectral composition [17–19]. Concerning economic viability, the use of LEDs is increasing rapidly in agriculture due to their huge capacity to save electrical energy. LEDs are more efficient in in vitro culture than white fluorescent light. It is stipulated that they have the specific wavelengths that fits plants exact need for morphogenic responses [20]. The wavelength a plant needs varies according to the species.

LEDs are a unique type of semiconductor diode that have several technical benefits over usual light sources for photosynthesis [21]. LEDs allow wavelengths to be matched to the plant photoreceptors to influence plant morphology and metabolic composition [22–24]. Plant pigments absorb red wavelengths (600–700 nm) efficiently, with the most efficient being660 nm, which is close to the chlorophyll absorption peak, whereas the blue region includes the visible spectrum (400–500 nm) [25]. It has also been reported that red light has a significant role in starch accumulation through photosynthesis [26] and blue light in chloroplast development, chlorophyll formation, and stomatal opening [27]. Red and blue light are the best to drive photosynthetic metabolism. Green wavelength effects are opposite those of red and blue wavelengths [28].

A number of in vitro studies have reported vigorous plant growth under LEDs. LED lights have been previously used for PLB organogenesis in *Cymbidium finlaysonianum* [29], *Dendrobiumkingianum* [30,31], hybrid *Cymbidium* [32,33], and plantlet regeneration in gerbera [34]. A series of studies have already been conducted to improve the tissue culture of *Phalaenopsis* using a number of factors. We previously used growth regulators and elicitors for the PLB regeneration of *Phalaenopsis* [35,36]. The effects of light spectral quality on the photosynthetic ability varied by plant species [37]. Many plant species do not respond well under a sole LED color, and this limitation can be overcome by combining different colors. On the other hand, many researchers have reported the long-term effect of growth retardants on in vitro growth and development [38,39]. Chlorocholine chloride (CCC: (2-chloroethyl) trimethyl-ammonium chloride) can be used to manipulate plant growth [40–42]; it inhibits gibberellic acid biosynthesis [43,44]. Gibberellic acid reduces adenosine diphosphate-glucose pyrophosphorylase (AGPase) activity, which is responsible for the reduction of starch synthesis [45]. Application of CCC can counteract this starch synthesis reduction by blocking gibberellic acid synthesis. CCC may have an impact on plant growth by altering the hormone content.

The purpose of this study was to determine the best CHO source and LED light combination for successful PLB regeneration of *Phalaenopsis* 'Fmk02010'. In addition, our goals was also to assess the impact of CCC priming in in vitro PLB propagation of *Phalaenopsis*.

2. Materials and Methods

2.1. Plant Materials and Culture Conditions

PLBs of *Phalaenopsis* 'Fmk02010' were multiplied in 2.2 gL^{-1} of Phytagel^{TM} (Sigma-Aldrich$^{®}$, Tokyo, Japan) solidified MS medium (modified) [46] at the Lab of Vegetable and Floricultural Science, Faculty of Agriculture and Marine Science, Kochi University, Japan. We added two major salts, ammonium nitrate (412.5 mgL^{-1}) and potassium nitrate (950.0 mgL^{-1}), to the MS medium for the modification. We excised single PLBs to use as explants. The pH was adjusted to 5.5–5.8 using 1 mM 2-(N-morpholino) ethanesulfonic acid sodium salt (MES-Na) prior to autoclaving. We used 30 mL of culture media in each 250-mL culture bottle (UM culture bottle: AsOne, Japan) and autoclaved at 121 °C for 15 min at 117.1 KPa.

2.2. CHO Sources and LED Lights

Sucrose, trehalose, and maltose (20 g/L) (Sigma-Aldrich®, Tokyo, Japan) were used as CHO sources before autoclaving. The PLBs for organogenesis were placed under fourteen different LED light sources with a control. These were: (1) control (C: white fluorescent light); (2) R (red LED); (3) G (green LED); (4) B (blue LED); (5) W (white LED); (6) RG (red → green LED); (7) RB (red → blue LED); (8) RW (red → white LED); (9) GB (green → blue LED); (10) GW (green → white LED); (11) BW (blue → white LED); (12) RGB (red → green → blue LED); (13) RBW (red → blue → white LED); (14) RGW (red → green → white LED); and (15) GBW (green → blue → white LED). All LED lamps were monochromic. We did not use two different monochromic LEDs together. A monochromic light supplemented with ≥1 monochromic light had a dissimilar light effect. For example, red LEDs supplemented with blue fluorescent were equivalent to cool-white fluorescent plus incandescent lamps [47]. The technique for the sole, double, and triple LED light combinations used in this study is shown in Scheme 1.

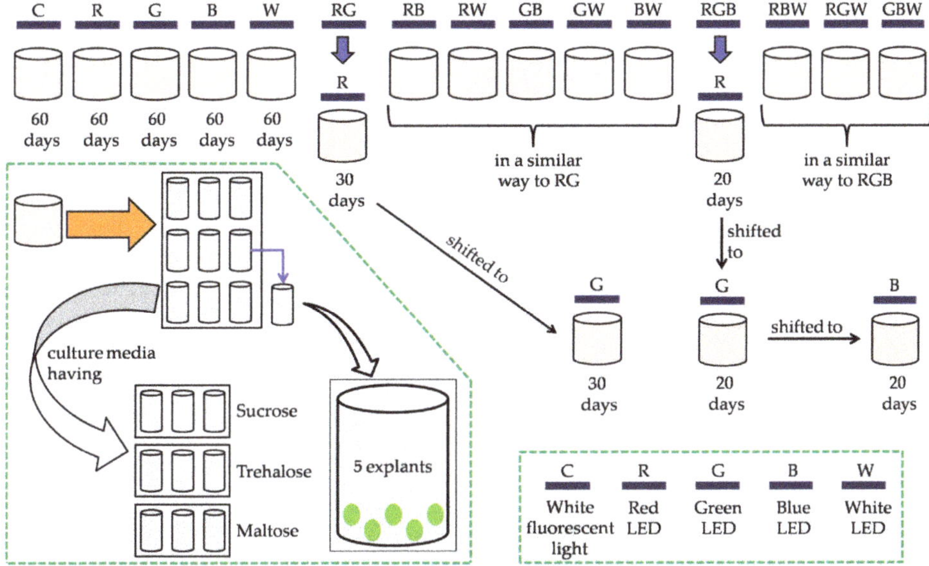

Scheme 1. Visualization of the experimental layout. C, control.

2.3. CCC Concentrations

Four different CCC concentrations with the control (BioReagent, Sigma-Aldrich®, Tokyo, Japan) were used. They were 0 (control), 0.01, 0.1, 1, and 10 mgL^{-1}. Sucrose was used in the culture media, while other culture conditions were similar as described in Section 2.1.

2.4. PLB Culture, Data Collection, and Data Analysis

Experiments were organized in a randomized complete block design. Each bottle contained five PLBs (with three replications). PLBs were cultured 60 days for the LED-carbon source experiments and 42 days for the CCC-treated PLBs. The explants were cultured at 25 ± 2 °C with a 16-h photoperiod with 54 μmol/m^{-2} s^{-1} of irradiance. The number of PLBs (including budding PLBs), shoots, and roots were counted (Figure 1a). The length of shoots and the fresh weight (FW) of PLBs were measured. The average numbers and percentages were calculated as follows.

■   Average number = Number of cultured explants with new PLBs/Total number of cultured explants

■   Percentage of PLB (%) = (Number of cultured explants with new PLBs/Total number of cultured explants) × 100

Data are presented as the mean ± the standard error (SE). One-way ANOVA was analyzed by Minitab®17 (Minitab Inc., Pennsylvania 16801-3008, USA, 2017) using Tukey's multiple comparisons test method with the 95% confidence interval.

3. Results

3.1. CHO Sources and LED Lights

RW-sucrose and the control did not significantly differ (Table 1). However, all other LED-sucrose combinations produced a significantly lower number of PLBs than the RW-sucrose, some significantly lower than the control as well. Within trehalose treatments, BW-trehalose performed well for the mean number of PLBs (36.33), which was closely followed by RBW-trehalose (35.13) (Table 1). Maltose showed the worst overall performance for PLB regeneration with all LED combinations (Table 1). PLBs under different LEDs showed statistically identical fresh weights for mediums with sucrose and most with trehalose. However, the medium with maltose showed significant differences among the different LEDs. Maximum mean fresh weight was found for RBW-trehalose (0.167 g), RBW-maltose (0.112 g), and RW-sucrose (0.109 g) (Table 1). The RBW-trehalose combination also produced a satisfactory number of PLBs (35.13) (Table 1). The CHO source-LED combinations with the first, second, and third highest values for the number of PLBs within each CHO source group (Figure 2) and fresh weight (Figure 3) are shown. Sucrose produced the highest and second highest numbers of PLBs, and trehalose was best for the fresh weight as the CHO source in the culture medium (Figures 2 and 3).

After 60 days of culture, some treatments tended to produce shoots. Trehalose had a greater tendency for shoot growth under LED lights except GW, RGB, RBW, RGW, and GBW (Table 2). There were no shoots for trehalose under white fluorescent light (control). Shoots were produced with W-sucrose and RGW-sucrose, as well as with RG-maltose. Root formation was not observed in any of the treatment combinations except trehalose-RG (number: 0.03; length: 0.03 cm; data are not shown).

3.2. CCC Concentrations

The number of PLBs, PLB formation rate, and fresh weight of *Phalaenopsis* 'Fmk02010' with different concentrations of CCC in the culture medium are shown in Table 3. The number of PLBs and fresh weight were significantly lower at 10 mgL^{-1}CCC. The maximum number of PLBs were produced in the culture medium treated with 0.01 mgL^{-1} of CCC. In this treatment, there was a 100% PLB formation rate. The PLB formation rates were 93.33%, 93.33%, 80.00%, and 33.33% at 0, 0.01, 1, and 10 mgL^{-1} of CCC, respectively. The maximum fresh weight was from the culture media having 0.01 mgL^{-1} of CCC, whereas the minimum fresh weight was found at 10 mgL^{-1} of CCC (Table 3). CCC at 0.01, 0.1, and 1 mg L^{-1} did not differ from the control values for number of PLBs or fresh weight. In the scatter plot (Figure 4), the relationship between PLB organogenesis and CCC concentration are shown. The R^2 of the correlation was high for both number of PLBs (R^2 = 0.915) and fresh weight (R^2 = 0.747) to the different CCC concentrations. There was a negative relationship in both cases, suggesting that the number of PLBs and fresh weight would decrease with increasing CCC concentration in the culture medium.

Table 1. Mean number of PLBs and fresh weight of *Phalaenopsis* 'Fmk02010' withdifferent CHO sources and LED lights.

Light [z]	Mean Number of PLBs			Fresh Weight (g)		
	Sucrose	Trehalose	Maltose	Sucrose	Trehalose	Maltose
Control	37.73 ± 4.40 [y] ab	28.13 ± 4.87 abcd	1.60 ± 0.53 b	0.098 ± 0.057 a	0.059 ± 0.033 ab	0.043 ± 0.023 ab
R	26.20 ± 3.38 bc	15.87 ± 2.89 bcde	2.73 ± 0.97 b	0.062 ± 0.042 a	0.059 ± 0.032 ab	0.033 ± 0.019 b
G	21.93 ± 3.76 bcd	8.87 ± 1.55 de	6.00 ± 1.35 b	0.066 ± 0.045 a	0.041 ± 0.022 ab	0.033 ± 0.018 b
B	20.47 ± 3.92 bcd	27.67 ± 3.21 abcd	2.27 ± 0.76 b	0.043 ± 0.024 a	0.137 ± 0.074 ab	0.017 ± 0.009 b
W	24.80 ± 4.49 bcd	22.80 ± 2.92 abcde	3.60 ± 0.67 b	0.071 ± 0.039 a	0.079 ± 0.042 ab	0.028 ± 0.017 b
RG	11.07 ± 3.97 cd	19.33 ± 3.44 abcde	4.47 ± 1.04 b	0.029 ± 0.022 a	0.063 ± 0.034 ab	0.043 ± 0.023 ab
RB	11.60 ± 3.11 cd	15.33 ± 5.55 bcde	1.47 ± 0.45 b	0.026 ± 0.020 a	0.091 ± 0.065 ab	0.024 ± 0.013 b
RW	54.13 ± 8.85 a	12.13 ± 2.82 cde	14.47 ± 3.39 a	0.109 ± 0.063 a	0.027 ± 0.020 b	0.066 ± 0.041 ab
GB	19.93 ± 4.83 bcd	22.73 ± 3.67 abcde	2.27 ± 1.03 b	0.061 ± 0.035 a	0.117 ± 0.062 ab	0.020 ± 0.011 b
GW	26.40 ± 3.60 bc	8.73 ± 2.19 de	4.80 ± 1.11 b	0.080 ± 00.047 a	0.047 ± 0.026 ab	0.051 ± 0.027 ab
BW	10.67 ± 4.11 cd	36.33 ± 5.08 a	3.80 ± 1.76 b	0.020 ± 0.020 a	0.129 ± 0.071 ab	0.045 ± 0.024 ab
RGB	5.47 ± 1.98 d	5.00 ± 1.70 e	8.40 ± 1.85 a	0.015 ± 0.015 a	0.066 ± 0.048 ab	0.079 ± 0.047 ab
RBW	13.53 ± 2.88 cd	35.13 ± 4.36 ab	13.73 ± 2.09 a	0.034 ± 0.024 a	0.167 ± 0.098 a	0.112 ± 0.068 a
RGW	19.07 ± 2.60 bcd	29.40 ± 4.45 abc	1.47 ± 0.35 b	0.030 ± 0.021 a	0.090 ± 0.048 ab	0.022 ± 0.013 b
GBW	12.20 ± 2.79 cd	32.00 ± 7.77 ab	4.73 ± 1.27 b	0.056 ± 0.038 a	0.088 ± 0.063 ab	0.030 ± 0.017 b

[z] Control (C: white fluorescent light); R (red LED); G (green LED); B (blue LED); W (white LED); RG (red → green LED); RB (red → blue LED); RW (red → white LED); GB (green → blue LED); GW (green → white LED); BW (blue → white LED); RGB (red → green → blue LED); RBW (red → blue → white LED); RGW (red → green → white LED); and GBW (green → blue → white LED). [y] Mean± SE values that do not share a letter are significantly different within each column, and those sharing a letter are statistically similar at $P \leq 0.05$.

Table 2. Shoot growth withdifferent CHO sources and LED lights during PLB organogenesis of *Phalaenopsis* 'Fmk02010'.

Light [z]	Number of Shoots			Mean Shoot Length		
	Sucrose	Trehalose	Maltose	Sucrose	Trehalose	Maltose
Control	0	0	0	0	0	0
R	0	0.38 ± 0.14	0	0	0.26 ± 0.10	0
G	0	0.38 ± 0.14	0	0	0.04 ± 0.02	0
B	0	0.75 ± 0.29	0	0	0.11 ± 0.05	0
W	0.13 ± 0.07	0.25 ± 0.09	0	0.05 ± 0.05	0.08 ± 0.03	0
RG	0	0.13 ± 0.07	0.13 ± 0.13	0	0.08 ± 0.04	0.03 ± 0.03
RB	0	0.25 ± 0.13	0	0	0.05 ± 0.03	0
RW	0	0.13 ± 0.07	0	0	0.03 ± 0.01	0
GB	0	0.13 ± 0.07	0	0	0.05 ± 0.03	0
GW	0	0	0	0	0	0
BW	0	0.38 ± 0.14	0	0	0.12 ± 0.04	0
RGB	0	0	0	0	0	0
RBW	0	0	0	0	0	0
RGW	0.25 ± 0.13	0	0	0.03 ± 0.03	0	0
GBW	0	0	0	0	0	0

[z] Control (C: white fluorescent light); R (red LED); G (green LED); B (blue LED); W (white LED); RG (red → green LED); RB (red → blue LED); RW (red → white LED); GB (green → blue LED); GW (green → white LED); BW (blue → white LED); RGB (red → green → blue LED); RBW (red → blue → white LED); RGW (red → green → white LED); and GBW (green → blue → white LED).

Table 3. Role of CCC concentrations forthe in vitro PLB production of *Phalaenopsis* 'Fmk02010'.

CCC (mgL^{-1})	Number of PLBs	PLB Formation (%)	Fresh Weight (g)
0	12.53 ± 1.71 [z] a	93.33	0.175 ± 0.028 a
0.01	15.67 ± 1.01 a	100.00	0.211 ± 0.018 a
0.1	13.73 ± 1.62 a	93.33	0.191 ± 0.022 a
1	11.07 ± 2.08 ab	80.00	0.182 ± 0.027 a
10	4.40 ± 1.74 b	33.33	0.049 ± 0.019 b

[z] Mean ± SE values that do not share a letter within each column are significantly different at $P \leq 0.05$.

Figure 1. PLB organogenesis of *Phalaenopsis*: (**a**) PLBs, shoots, and roots, (**b**) RW-sucrose; and (**c**) BW-trehalose.

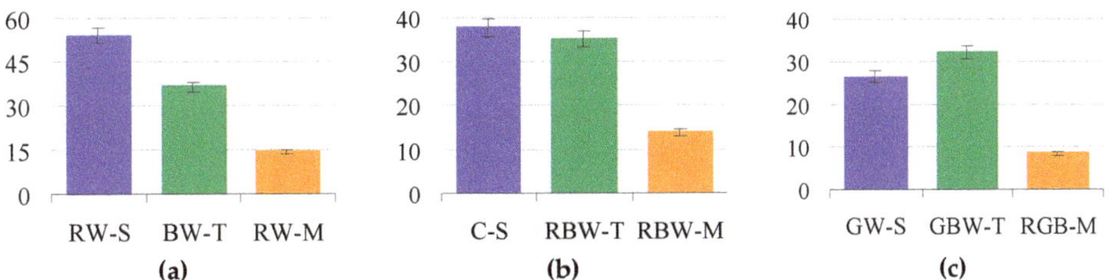

Figure 2. Comparison of PLB production for the (**a**) highest (**b**) second highest and (**c**) third highest CHO source –LED combination within each CHO source. The X-axis the treatment combination, the Y-axis represents the number of PLBs, and the error bar represents the 95% confidence intervals. See text for meanings of abbreviations.

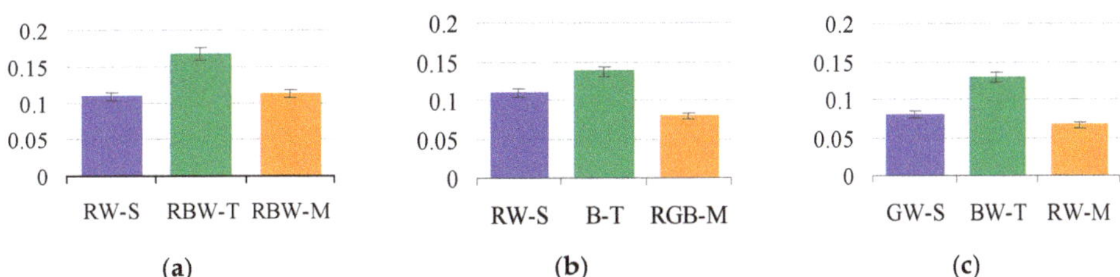

Figure 3. Comparison of fresh weight for the (**a**) highest (**b**) second highest and (**c**) third highest within CHO source –LED combination each CHO source. The X-axis the treatment combination, the Y-axis represents the number of PLBs, and the error bar represents the 95% confidence intervals. See text for meanings of abbreviations.

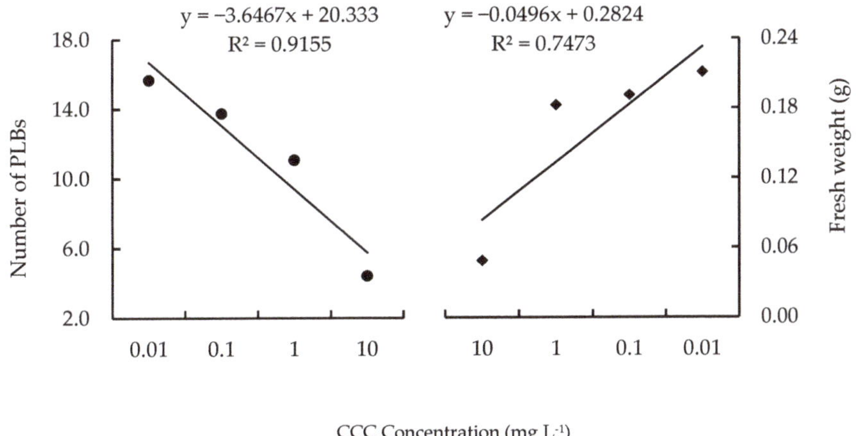

Figure 4. Correlation between the number of PLBs (left) and fresh weight (right) with CCC concentrations of the culture medium. Mean data were used for these analyses.

4. Discussion

Among the combinations, RW-sucrose produced the maximum number of PLBs, but the fresh weight not the highest (Table 1). On the other hand, the BW-trehalose combination produced comparatively fewer PLBs than that of RW-sucrose, but the fresh weight was higher. Sucrose was the best CHO source for number of PLBs (Figure 2), whereas trehalose was the best regarding the fresh weight (Figure 3). Both number of PLBs and fresh weight are very important for successful and healthy PLB regeneration. Using trehalose in culture media was more effective than sucrose for friable callus formation in *Phalaenopsis* [48]. RW-sucrose, BW-trehalose, and RBW-trehalose combinations may be better for PLB organogenesis of *Phalaenopsis* considering both the number of PLBs and fresh weight. These three combinations were cultured in the white LED on the last 20 or 30 days; the first 20 or 30 days they were cultured under red, or blue, or red and blue LEDs. Results suggested that a white LED was important for rapid and healthy PLB growth, because the plant may have the ability to produce more chlorophyll under white light [49]. PLBs cultured under red LEDs for the early period showed a tendency to generate more PLBs. Plant growth was fragile under red light [50,51], and it stimulated endogenous gibberellins that cause cell proliferation and mitosis [52]. Red light increased multiplication rate [53], and our study also confirmed the increased multiplication of PLBs under red light. The red wavelengths (between 600 and 700 nm) were absorbed by plant pigments [23]. Hormones responsible for inflorescence formation, inflorescence elongation, and bud breakage were stimulated by red light [54]. PLBs cultured under blue LEDs in the early period produced a higher fresh weight. Trehalose-BW produced the maximum fresh weight of PLBs, and blue LEDs robustly encouraged PLB growth. Tanaka et al. [55] found blue LEDs to be effective for PLB formation in *Phalaenopsis*.

LEDs are an effective light source for plants [52,56], and light quality plays apart in the vital function of photosynthesis. The mechanism is that in which light is absorbed by chlorophyll [57]. Blue light plays an important role in chlorophyll biosynthesis [58–60] that may affect both the number of PLBs and fresh weight with a white LED. Chlorophyll contents are correlated with plant species or cultivar when grown under different light qualities [61]. Anuchai and Hsieh [62] found significantly higher chlorophyll (both a and b) and carotenoid content under blue light in *Phalaenopsis*. They also found a higher number of stems, fresh weight, and leaf length under red LEDs and higher RuBisCO enzyme activity. PLBs cultured under red LEDs in the early period, and then shifted to blue LEDs and white LEDs, showed significantly better results both for number of PLBs and fresh weight. The results suggested that red, or blue, or red and blue LEDs should be used initially, and then shifted to white LEDs for successful and healthier PLB regeneration of *Phalaenopsis* 'Fmk02010', but their effects also depend on the CHO sources in the culture medium. In our study, trehalose was better for available CHO for PLB organogenesis with BW LEDs (i.e., 30 days under a blue LED → 30 days under a white LED). Similarly, sucrose was better for available CHO under RW LEDs (i.e., 30 days under a red LED → 30 days under a white LED). Conversely, BW-trehalose produced the second largest number of shoots (Table 2). LEDs have been successfully applied in vitro in various plant species [20,29–33]. The ideal light stipulation for each plant species is unique. The response to the spectral composition of one plant species in vitro may not be similar for another plant species [63].

In our previous study, we used a number of growth regulators for PLB regeneration [35,36], so included the growth retardant CCC in the current study. The concentration of CCC played an important role in PLB organogenesis of *Phalaenopsis* "Fmk02010" 'Fmk02010'. Growth retardants were extensively used in vivo to improve floricultural characteristics, especially to control plant height. Application of CCC seemed to be effective with a very low concentration. PLB formation was very sensitive to a high concentration (>0.01 mgL^{-1}). Plant growth retardants like CCC could improve carbohydrate accumulation by increasing photosynthetic capacity and altering endogenous hormones [64,65]. CCC treatment can promote nutrient uptake, water balance, and protein synthesis in growing organs [66]. An increasing concentration of CCC resulted in a decreased number and percentage of PLB formation. The addition of CCC to the in vitro medium enhanced tuberization [67–69]. We found an effect of a higher concentration CCC on PLB organogenesis in culture media through the investigation of the relationship between CCC with PLB organogenesis. Similar relationships have also been studied previously [70–72]. CCC is an anti-gibberellin growth regulator that inhibits an early step in gibberellic acid (GA) biosynthesis [43]. Treatments with CCC counteract the reduction in starch synthesis by blocking GA synthesis. Plant growth retardants like CCC and paclobutrazol are able to inhibit gibberellin biosynthesis or action [73,74], and can control excessive vegetative growth [75,76] which ultimately increases quality attributes such as dry matter content [77]. CCC treatment is mostly effective for tuberous and bulbous pants. PLBs are tuber-like bodies. We observed that the addition of the growth retardant CCC had no effect on PLB formation or fresh weight except for a reduction at the highest concentration (Table 3).

5. Conclusions

Sucrose and trehalose can be used as excellent CHO sources in the culture media for PLB regeneration of *Phalaenopsis*. RW-sucrose was the best combination to produce the maximum number of PLBs. However, the combination of BW-trehalose also produced a large number with healthier PLBs; it also had a tendency to produce a greater number of shoots that would need immediate subculture for future preservation. RBW-trehalose generated a satisfactory number of PLBs with a higher fresh weight and did not generate any shoots. An excessive concentration of CCC (10 mgL^{-1}) caused enormous reduction in the number of PLBs, the percentage of PLB formation, as well as fresh weight.

Author Contributions: H.M. (Hasan Mehraj) conceptualized and executed the study, as well as prepared the original draft of the manuscript; M.M.A. and H.M. (Hasan Mehbub) were responsible for data collection, compilation, and formal analysis; the manuscript was edited by S.U.H.

Funding: This research received no external funding.

Acknowledgments: We are especially thankful to the lab members who helped in the PLB subculture. We are grateful to Kazuhiko Shimasaki (Kochi University, Japan) for his technical guidance and lab support.

Conflicts of Interest: The authors declare that there is no conflict of interest.

References

1. Arditti, J.; Ernst, R. *Micropropagation of Orchids*; Wiley: New York, NY, USA, 1993; pp. 1–682.
2. Sheelavanthmath, S.S.; Murthy, H.N.; Hema, B.P.; Hahn, E.J.; Paek, K.Y. High frequency of protocorm like bodies (PLBs) induction and plant regeneration from protocorm and leaf sections of *Aerides Crispum*. *Sci. Hortic.* **2005**, *106*, 395–401. [CrossRef]
3. Tokuhara, K.; Mii, M. Micropropagation of *Phalaenopsis* and *Doritaenopsis* by shoot tips of flower stalk buds. *Plant Cell Rep.* **1993**, *13*, 7–11. [CrossRef] [PubMed]
4. Ichihashi, S. Micropropagation of *Phalaenopsis* through the culture of lateral buds from young flower stalks. *Lindleyana* **1992**, *7*, 208–215.
5. Tanaka, M.; Senda, Y.; Hasegawa, A. Plantlet formation in root-tip culture in *Phalaenopsis*. *Am. Orchid Soc. Bull.* **1976**, *45*, 1022–1024.
6. Park, S.Y.; Murthy, H.N.; Paek, K.Y. Rapid propagation of Phalaenopsis from floral stalk-derived leaves. *In Vitro Cell. Dev. Biol. Plant* **2002**, *38*, 168–172. [CrossRef]
7. Tokuhara, K.; Mii, M. Induction of embryogenic callus and cell suspension culture from shoot tips excised from flower stalk buds in *Phalaenopsis* (Orchidaceae). *In Vitro Cell. Dev. Biol. Plant* **2001**, *37*, 457–461. [CrossRef]
8. Tokuhara, K.; Mii, M. Highly-efficient somatic embryogenesis from cell suspension cultures of *Phalaenopsis* orchids by adjusting carbohydrate sources. *In Vitro Cell. Dev. Biol. Plant* **2003**, *39*, 635–639. [CrossRef]
9. Chen, Y.C.; Chang, C.; Chang, W.C. A reliable protocol for plant regeneration from callus culture of *Phalaenopsis*. *In Vitro Cell. Dev. Biol. Plant* **2000**, *36*, 420–423. [CrossRef]
10. Park, S.Y.; Yeung, E.C.; Chakrabarty, D.C.; Paek, K.Y. An efficient direct induction of protocorm-like bodies from leaf sub epidermal cells of *Doritaenopsis* hybrid using thin-section culture. *Plant Cell Rep.* **2002**, *21*, 46–51. [CrossRef]
11. Bustam, S.; Sinniah, U.R.; Swamy, M.K. Simple and efficient in vitro method of storing *Dendrobium* sw Shavin White protocorm like bodies (PLBs). *Bangladesh J. Bot.* **2017**, *46*, 439–449.
12. Murdad, R.; Latip, M.A.; Aziz, Z.A.; Ripin, R. Effects of carbon source and potato homogenate on in vitro growth and development of Sabah's endangered orchid: *Phalaenopsis gigantea*. *AsPac J. Mol. Biol. Biotechnol.* **2010**, *18*, 199–202.

13. Al-Khateeb, A.A. Regulation of in vitro bud formation of date palm (*Phoenix dactylifera* L.) cv. Khanezi by different carbon sources. *Bioresour. Technol.* **2008**, *99*, 6550–6555. [CrossRef] [PubMed]

14. Liu, T.H.A.; Lin, J.J.; Wu, R.Y. The effects of using trehalose as a carbon source on the proliferation of *Phalaenopsis* and *Doritaenopsis* protocorm-like-bodies. *Plant Cell Tissue Org. Cult.* **2006**, *86*, 125–129. [CrossRef]

15. Akter, S.; Nasiruddin, K.M.; Khaldun, A.B.M. Organogenesis of *Dendrobium* orchid using traditional media and organic extracts. *J. Agric. Rural Dev.* **2007**, *5*, 30–35. [CrossRef]

16. Taiz, L.; Zeiger, E. *Plant Physiology*, 1st ed.; Benjamin-Cumings Publishing Co.: New York, NY, USA, 1991.

17. Agarwal, A.; Gupta, S.D. Impact of light-emitting diodes (LEDs) and its potential on plant growth and development in controlled environment plant production system. *Curr. Biotechnol.* **2016**, *5*, 28–43. [CrossRef]

18. Bello-Bello, J.J.; Pérez-Sato, J.A.; Cruz-Cruz, C.A.; Martínez-Estrada, E. Light-emitting diodes: Progress in plant micropropagation. In *Chlorophyll*; Jacob-Lopes, E., Ed.; InTech: London, UK, 2017; pp. 747–1216.

19. Shukla, M.R.; Singh, A.S.; Piunno, K.; Saxena, P.K.; Jones, A.M.P. Application of 3D printing to prototype and develop novel plant tissue culture systems. *Plant Methods* **2017**, *13*, 6–15. [CrossRef] [PubMed]

20. Gupta, S.D.; Jatothu, B. Fundamentals and applications of light-emitting diodes (LEDs) in in vitro plant growth and morphogenesis. *Plant Biotechnol. Rep.* **2013**, *7*, 211–220. [CrossRef]

21. Mitchell, C.A.; Both, A.J.; Bourget, C.M.; Burr, J.F.; Kubota, C.; Lopez, R.G.; Morrow, R.C.; Runkle, E.S. LEDs: The future of greenhouse lighting. *Chron. Hortic.* **2012**, *52*, 6–10.

22. Bourget, C.M. An introduction to light-emitting diodes. *HortScience* **2008**, *43*, 1944–1946. [CrossRef]

23. Massa, G.D.; Kim, H.H.; Wheeler, R.M.; Mitchell, C.A. Plant productivity in response to LED lighting. *HortScience* **2008**, *43*, 1951–1956. [CrossRef]

24. Morrow, R.C. LED lighting in horticulture. *HortScience* **2008**, *43*, 1947–1950. [CrossRef]

25. Matioc-Precup, M.M.; Cachiţă-Cosma, D. The germination and growth of *Brassica oleracea* L. var. *capitata* f. rubra plantlets under the influence of colored light of different provenance. *Studia Univ. Vasile Goldis Arad Ser. Stiintele Vietii* **2012**, *22*, 193–202.

26. Saebo, A.; Krekling, T.; Appelgren, M. Light quality affects photosynthesis and leaf anatomy of birch plantlets in vitro. *Plant Cell Tissue Org. Cult.* **1995**, *41*, 177–185. [CrossRef]

27. Senger, H. The effect of blue light on plants and microorganisms. *Phytochem. Photobiol.* **1982**, *35*, 911–920. [CrossRef]

28. Folta, K.M.; Maruhnich, A.S. Green light: A signal to slow down or stop. *J. Expt. Bot.* **2007**, *58*, 3099–3111. [CrossRef] [PubMed]

29. Nahar, S.J.; Haque, S.M.; Shimasaki, K. Organogenesis of *Cymbidium finlaysonianum* under different sources of lights. *Am. Eurasian J. Agric. Environ. Sci.* **2015**, *15*, 2095–2101.

30. Habiba, S.U.; Shimasaki, K.; Ahasan, M.M.; Alam, M.M. Effects of different light quality on growth and development of protocorm-like bodies (PLBs) in *Dendrobium kingianum* cultured in vitro. *Bangladesh Res. Public J.* **2014**, *10*, 223–227.

31. Habiba, S.U.; Shimasaki, K.; Ahasan, M.M.; Alam, M.M. Effect of 6-Benzylaminopurine (BA) and Hyaluronic Acid (HA) under white light emitting diode (LED) on organogenesis in protocorm-like bodies (PLBs) of *Dendrobium kingianum*. *Am. Eurasian J. Agric. Environ. Sci.* **2014**, *14*, 605–609.

32. Kamal, M.M.; Shimasaki, K.; Akhter, N. Effect of light emitting diode (LED) and N-Acetyleglucoseamine (NAG) on organogenesis in protocorm-like-bodies (PLBs) of *Cymbidium* hybrid cultured In Vitro. *Plant Tissue Cult. Biotech.* **2014**, *24*, 273–277. [CrossRef]

33. Teixeira da Silva, J.A. The response of protocorm-like bodies of nine hybrid Cymbidium cultivars to light-emitting diodes. *Environ. Exp. Biol.* **2014**, *12*, 155–159.

34. Wang, Z.; Li, G.; He, S.; Teixeira da Silva, J.A.; Tanaka, M. Effect of cold cathode fluorescent lamps on growth of *Gerbera jamesonii* plantlets in vitro. *Sci. Hortic.* **2011**, *130*, 482–484. [CrossRef]

35. Sultana, K.S.; Mustafa, K.H.; Mehedi, K.H.; Sultana, S.; Mehraj, H.; Shimasaki, K.; Habiba, S.U. Effect of Hyaluronic Acid (HA) on organogenesis in protocorm-like-bodies (PLBs) of *Phalaenopsis* 'Fmk02010' cultured in vitro. *Am. Eurasian J. Agric. Environ. Sci.* **2015**, *15*, 1721–1724.

36. Sultana, K.S.; Mustafa, K.H.; Mehedi, K.H.; Sultana, S.; Mehraj, H.; Shimasaki, K.; Habiba, S.U. Effect of two elicitors on organogenesis in protocorm-like bodies (PLBs) of *Phalaenopsis* 'Fmk02010' cultured in vitro. *World Appl. Sci. J.* **2015**, *33*, 1528–1532.

37. Silva Batista, D.; Felipe, S.; Heitor, S.; Dulcineia Silva, T.; Motta de Castro, K.; Mamedes-Rodrigues, T.C.; Amaral Miranda, N.; Ríos-Ríos, A.M.; Vidal Faria, D.; Alexandre Fortini, E.; et al. Light quality in plant tissue culture: Does it matter? *In Vitro Cell. Dev. Biol. Plant* **2018**, *54*, 195–215. [CrossRef]

38. Ruter, J.M. Growth and landscape establishment of *Pyracantha* and *Juniperus* after application of paclobutrazol. *HortScience* **1994**, *29*, 1318–1320. [CrossRef]

39. Kozak, D.; Grodek, J. The consequent effect of growth retardants on the growth and development of *Tibouchina urvilleana* Cogn. shoots in vitro. *Acta Sci. Pol. Hortorum Cultus* **2005**, *4*, 123–128.

40. Sharma, N.; Kaur, N.; Gupta, A.K. Effects of gibberellic acid and chlorocholine chloride on tuberization and growth of potato (*Solanumtuberosum* L.). *J. Sci. Food Agric.* **1998**, *78*, 466–470. [CrossRef]

41. Sharma, N.; Kaur, N.; Gupta, A.K. Effect of chlorocholine chloride sprays on the carbohydrate composition and activities of sucrose metabolising enzymes in potato (*Solanum tuberosum* L.). *Plant Growth Regul.* **1998**, *26*, 97–103. [CrossRef]

42. Berova, M.; Zlatev, Z. Physiological response and yield of paclobutrazol treated tomato plants (*Lycopersicon esculentum* Mill.). *Plant Growth Regul.* **2000**, *30*, 117–123. [CrossRef]

43. Wang, H.; Li, H.; Liu, F.; Xiao, L. Chlorocholine chloride application effects on photosynthetic capacity and photoassimilates partitioning in potato (*Solanum tuberosum* L.). *Sci. Hortic.* **2009**, *119*, 113–116. [CrossRef]

44. Wang, H.Q.; Xiao, L.T. Effects of chlorocholine chloride on phytohormones and photosynthetic characteristics in potato (*Solanum tuberosum* L.). *J. Plant Growth Regul.* **2009**, *28*, 21–27. [CrossRef]

45. Mares, D.J.; Marschner, H.; Krauss, A. Effect of gibberellic acid on growth and carbohydrate metabolism of developing tubers of potato (*Solanum tuberosum* L.). *Physiol. Plant.* **1981**, *52*, 267–274. [CrossRef]

46. Shimasaki, K.; Uemoto, S. Micropropagation of a terrestrial *Cymbidium* species using rhizomes developed from seeds and pseudobulbs. *Plant Cell Tissue Org. Cult.* **1990**, *22*, 237–244. [CrossRef]

47. Bula, R.J.; Morrow, R.C.; Tibbitts, T.W.; Barta, D.J.; Ignatius, R.W.; Martin, T.S. Light emitting diodes as a radiation source for plants. *HortScience* **1991**, *26*, 203–205. [CrossRef] [PubMed]

48. Duran, R.E.; Coskun, Y. In vitro culture response of Phalaenopsis orchid by different carbon sources. *J. Biotechnol.* **2015**, *208*, S107. [CrossRef]

49. Bello-Bello, J.J.; Martínez-Estrada, E.; Caamal-Velázquez, J.H.; Morales-Ramos, V. Effect of LED light quality on in vitro shoot proliferation and growth of vanilla (*Vanilla planifolia* Andrews). *Afr. J. Biotechnol.* **2016**, *15*, 272–277. [CrossRef]

50. Heo, J.W.; Shin, K.S.; Kim, S.K.; Paek, K.Y. Light quality affects in vitro growth of grape 'Teleki 5BB7'. *J. Plant Biol.* **2006**, *49*, 276–280. [CrossRef]

51. Moon, H.K.; Park, S.Y.; Kim, Y.W.; Kim, C.S. Growth of Tsuru-rindo (*Tripterospermum japonicum*) cultured in vitro under various sources of light-emitting diode (LED) irradiation. *J. Plant Biol.* **2006**, *49*, 174–179. [CrossRef]

52. Manivannan, A.; Soundararajan, P.; Halimah, N.; Ko, C.H. Blue LED light enhances growth, phytochemical contents, and antioxidant enzyme activities of *Rehmannia glutinosa* cultured in vitro. *Hortic. Environ. Biotechnol.* **2015**, *56*, 105–113. [CrossRef]

53. Mengxi, L.; Zhigang, X.; Yang, Y.; Yijie, F. Effects of different spectral lights on *Oncidium* PLBs induction, proliferation, and plant regeneration. *Plant Cell Tissue Org. Cult.* **2011**, *106*, 1–10. [CrossRef]

54. Dueck, T.A.; Trouwborst, G.; Hogewoning, S.; Meinen, E. Can a high red: Far red ratio replace temperature induced inflorescence development in Phalaenopsis? *Environ. Exp. Bot.* **2016**, *121*, 139–144. [CrossRef]

55. Tanaka, M.; Watanabe, T.; Giang, D.T.; Tanaka, M.; Takamura, T.; Watanabe, H. Morphogenesis in the PLB segments of *Phalaenopsis* cultured under LED irradiation system (Abstract only). *J. Jpn. Soc. Hortic. Sci.* **2001**, *70* (Suppl. 1), 306.

56. Lin, Y.; Li, J.; Li, B.; He, T.; Chun, Z. Effects of light quality on growth and development of protocorm-like bodies of *Dendrobium officinale* in vitro. *Plant Cell Tissue Org. Cult.* **2011**, *105*, 329–335. [CrossRef]

57. Topchiy, N.M.; Sytnik, S.K.; Syvash, O.O.; Zolotareva, O.K. The effect of additional red irradiation on the photosynthetic apparatus of *Pisumsativum*. *Photosynthetica* **2005**, *43*, 451–456. [CrossRef]

58. Li, H.M.; Tang, C.M.; Xu, Z.G.; Liu, X.Y.; Han, X.L. Effects of different light sources on the growth of non-heading Chinese cabbage (*Brassica campestris* L.). *J. Agric. Sci.* **2012**, *4*, 262–273. [CrossRef]

59. Li, H.M.; Xu, Z.G.; Tang, C.M. Effect of light emitting diodes on growth and morphogenesis of upland cotton (*Gossypium hirsutum* L.) plantlets in vitro. *Plant Cell Tissue Org. Cult.* **2010**, *103*, 155–163. [CrossRef]

60. Kurilcik, A.; Miklusyte-Canova, R.; Dapkuniene, S.; Zilinskaite, S.; Kurilcik, G.; Tamulaitis, G.; Duchovskis, P.; Zukauskas, A. In vitro culture of chrysanthemum plantlets using light-emitting diodes. *Cent. Eur. J. Biol.* **2008**, *3*, 161–167. [CrossRef]

61. Li, H.M.; Tang, C.; Xu, Z. The effects of different light qualities on rapeseed (*Brassica napus* L.) plantlet growth and morphogenesis in vitro. *Sci. Hortic.* **2013**, *150*, 117–124. [CrossRef]

62. Anuchai, J.; Hsieh, C.H. Effect of change in light quality on physiological transformation of in vitro Phalaenopsis 'Fortune Saltzman' seedlings during the growth period. *Hortic. J.* **2017**, *86*, 395–402. [CrossRef]

63. Gupta, S.D.; Agarwal, A. Influence of LED lighting on in vitro plant regeneration and associated cellular redox balance. In *Light Emitting Diodes for Agriculture*; Dutta Gupta, S., Ed.; Springer: Singapore, 2017.

64. Zheng, R.; Wu, Y.; Xia, Y. Chlorocholine chloride and paclobutrazol treatments promote carbohydrate accumulation in bulbs of *Lilium* Oriental hybrids 'Sorbonne'. *J. Zhejiang Univ. Sci. B* **2012**, *13*, 136–144. [CrossRef]

65. Tezuka, T.; Takahara, C.; Yamamoto, Y. Aspects regarding the action of CCC in hollyhock plants. *J. Exp. Bot.* **1989**, *40*, 689–692. [CrossRef]

66. Grossmann, K. Plant growth retardants as tools in physiological research. *Physiol. Plant.* **1990**, *78*, 640–648. [CrossRef]

67. Hussain, I.; Chaudhry, Z.; Muhammad, A.; Aasghar, R.; Naqvi, S.M.S.; Rashid, H. Effect of chlorocholine chloride, sucrose and BAP on in vitro tuberization in potato (*Solanum tuberosum* L. cv. Cardinal). *Pak. J. Bot.* **2006**, *38*, 275–282.

68. Ray, A.; Bhattacharya, S. An improved micropropagation of *Eclipta alba* by in vitro priming with chlorocholine chloride. *Plant Cell Tissue Org. Cult.* **2007**, *92*, 315–319. [CrossRef]

69. Zakaria, M.; Hossain, M.M.; KhalequeMian, M.A.; Hossain, T.; Uddin, M.Z. In vitro tuberization of potato influenced by benzyl adenine and chloro choline chloride. *Bangladesh J. Agric. Res.* **2008**, *33*, 419–425. [CrossRef]

70. Coleman, W.K.; Donnelly, D.J.; Coleman, S.E. Potato microtubers as research tools: A review. *Am. J. Potato Res.* **2001**, *78*, 47–55. [CrossRef]

71. El-Sawy, A.; Bekheet, S.; Aly, U.I. Morphological and molecular characterization of potato microtubers production on coumarin inducing medium. *J. Agric. Biol.* **2007**, *9*, 675–680.

72. Gopal, J.; Chamail, A.; Sarkar, D. In vitro production of microtubers for conservation of potato germplasm: Effect of genotype, abscisic acid, and sucrose. *In Vitro Cell. Dev. Biol. Plant* **2004**, *40*, 485–490. [CrossRef]

73. Fletcher, A.; Gilley, A.; Sankhla, N.; Davies, T. Triazoles as plant growth regulators and stress protectants. *Hortic. Rev.* **1999**, *24*, 55–138. [CrossRef]

74. Mansuroglu, S.; Karaguzel, O.; Ortacesme, V.; Sayan, M. Effect of paclobutrazol on flowering, leaf and flower colour of *Consolida orientalis*. *Pak. J. Bot.* **2009**, *41*, 2323–2332.

75. Rademacher, W. Growth retardants: Effects on gibberellin biosynthesis and other metabolic pathways. *Annu. Rev. Plant Physiol. Plant Mol. Biol.* **2000**, *51*, 501–531. [CrossRef] [PubMed]

76. Ghosh, A.; Chikara, J.; Chaudhary, D.; Prakash, A.; Boricha, G.; Zala, A. Paclobutrazol arrests vegetative growth and unveils unexpressed yield potential of *Jatropha curcas*. *J. Plant Growth Regul.* **2010**, *29*, 307–315. [CrossRef]

77. Tsegaw, T.; Hammes, P. Growth responses of potato (*Solanum tuberosum*) grown in a hot tropical lowland to applied paclobutrazol: 2. Tuber attributes. *New Z. J. Crop Hortic. Sci.* **2005**, *33*, 43–51. [CrossRef]

 *horticulturae*

Communication

Aeroponic Cloning of *Capsicum* spp.

Angel R. Del Valle-Echevarria [1], Michael B. Kantar [1,*], Julianne Branca [2], Sarah Moore [1], Matthew K. Frederiksen [2], Landon Hagen [2], Tanveer Hussain [3] and David J. Baumler [2,4,5,*]

1   Department of Tropical Plant & Soil Sciences, University of Hawaii at Manoa, St. John Plant Science Laboratory, Room 102, 3190 Maile Way, Honolulu, HI 96822, USA; angeldve@hawaii.edu (A.R.D.V.-E.); moor1579@umn.edu (S.M.)
2   Department of Food Science and Nutrition, University of Minnesota-Twin Cities, 225 Food Science and Nutrition, 1334 Eckles Ave, St. Paul, MN 55108, USA; branc085@umn.edu (J.B.); frede345@umn.edu (M.K.F.); hagen626@umn.edu (L.H.)
3   Institute of Horticultural Sciences, University of Agriculture, Faisalabad 38040, Punjab, Pakistan; ch.tanver@gmail.com
4   Microbial and Plant Genome Institute, University of Minnesota-Twin Cities, 225 Food Science and Nutrition, 1334 Eckles Ave, St. Paul, MN 55108, USA
5   Biotechnology Institute, University of Minnesota-Twin Cities, 140 Gortner Lab, 1479 Gortner Ave., St Paul, MN 55108, USA
*   Correspondence: mbkantar@hawaii.edu (M.B.K.); dbaumler@umn.edu (D.J.B.)

Received: 9 March 2019; Accepted: 10 April 2019; Published: 16 April 2019

Abstract: Aeroponic cloning is a great strategy to maintain desired genotypes by generating a whole new plant from cuttings. While this propagation technique has been demonstrated for tomatoes (*Solanum lycopersicum*) and potatoes (*Solanum tuberosum*), no protocol has been developed for peppers (*Capsicum* spp.). The ability to clonally propagate different *Capsicum* holds promise for domestic and industrial growing operations since elite cultivars with desirable traits (e.g., high capsaicin levels, nutrient content, and striped fruit) can be perpetuated without the need of planning a nursery. We tested six *Capsicum* species for their feasibility of aeroponic cloning by stem cuttings. All domestic species were successfully regenerated under aeroponic conditions but not for *Capsicum eximium*, a wild species. Of the species analyzed, *Capsicum annuum* peppers had the fastest node formation (11.6 +/− 0.89 days, $P \leq 0.01$) and obtained a larger volume of roots ($P \leq 0.01$) after node formation as compared to *C. baccatum*, *C. frutescens*, and *C. pubescens*. This study presents a cost-effective strategy to clonally propagate peppers for personal, industrial, and conservation purposes.

Keywords: pepper; propagation; domestic; wild

---

1. Introduction

The genus *Capsicum* consists of more than 30 species, five of which are the result of domestication dating back to 6000 BC: *C. annuum*, *C. baccatum*, *C. chinense*, *C. frutescens*, and *C. pubescens* [1,2]. Commercially-cultivated pepper is an important vegetable and spice consumed daily by nearly a quarter of the world's population [3,4]. Domesticated in South-central Mexico (*C. annuum*; [5]) or Peru (*C. baccatum* and *C. pubescens*; [6]), peppers spread rapidly across the world after European contact due to their multitude of culinary uses, high nutritional content, and unique chemistry (e.g., capsaicin) [4]. Peppers are grown extensively worldwide with about 32 million metric tonnes produced in 2014 [7]. In 2016, the United States pepper production was valued at approximately 700 million USD [8]. In addition to the fresh market, there are many other high value products such as red pepper, chili flakes, and hot sauce that have distinct niche markets [9,10]. The most common pepper species in cuisine often depends on local cultural traditions. For example, in northerly temperate latitudes sweet bell peppers (*C. annuum*) are favored, while in many tropical regions fiery types

(*C. chinense*) are preferred [11]. Furthermore, *Capsicum* spp. cultivars produce varying levels of capsaicin, provitamin A, vitamin C, and folate, which has sparked interest in developing varieties to combat human malnutrition [12,13].

While pungency is an important characteristic in peppers [14], there are several roadblocks to propagating pungent plants. Often pungent varieties have poor and inconsistent germination, require longer time to germinate, and are expensive (more than 1 USD per seed). Additionally, many boutique growers make their own novel hybrids, both within and between species, and select for specific ideotypes [15].

Traditionally, pepper seed production is achieved by the self-pollination of promising plants to develop true breeding lines. In recent years, there has been an increase in the use of hybrid cultivars because of their higher yields and often have unique characteristics. However, seeds from hybrid plants cannot be saved for the next generation since self-progeny would show segregation of desirable traits. Therefore, a clonal propagation protocol could be implemented to enhance profitability of these novel germplasm types.

Clonally propagated pepper hybrids would provide an effective way to immortalize favored hybrid genotypes. Additionally, cloning protocols could be used to preserve pepper lineages that exhibit medicinal qualities or that are threatened by extinction [16]. While a major domestication trait across the plant kingdom has been a reduction in seed dormancy [17], which leads to easier use of plants in farming systems, it is unclear if this has also been selection for an increase in the ability of domestic species to be cloned, which would lead to a similar ease of use in cultivated systems.

Tissue culture is a method that is often used to clonally propagate recalcitrant, unique, or valuable samples by taking seed, embryos, or somatic tissue and placing it on a nutrient growth media. Tissue culture based methods have been developed to propagate successfully various *Capsicum* spp. [18] in laboratory conditions. However, tissue culture has limited utility for boutique or home growers as they do not have access to the sterile facilities or the necessary materials and supplies to carry out this task. Aeroponic cloning promises to be a great technique that can be used by pepper enthusiasts since it consists of soilless culture that allows for plant growth in a controlled environment, such as a greenhouse or growth chamber that can be readily available [19]. However, peppers have not been evaluated using this technique. The objective of this study was to evaluate the suitability of an aeroponic cloning protocol by using five domestic *Capsicum* spp. as well as one closely related wild species. Two key phenotypes that measure the success of this process are: time to node formation to judge the start of active growth (analogous to germination) and biomass accumulation, which is an important marker of plant vigor and a good predictor of future yield.

2. Materials and Methods

2.1. Propagation of Mother Plants

Pepper types were sourced from various heirloom seed producers across North America and represent six different species (unless otherwise noted; Table 1). Seeds from each pepper type were planted 0.635 cm deep using Master Garden premium potting mix (1.25N/0.21P/0.30K) (Premier Tech Horticulture, Ltd., Olds, AB, Canada) and sown in covered 32 cell deep flats for 7–14 days. Once plants emerged from soil, covers were removed from flats and plants were grown indoors for 18 h of light and 6 hours of a dark photoperiod under Sunblaze T5 fluorescent lights 5000 lumens (Sunlight Supply, Inc., Vancouver, WA, USA). All plants were fertilized with MotherPlant Nutrients (0.2N/0.3P/0.2K) (Hydrodynamics International, Lansing, MI, USA) following the manufacturer's recommendations.

Table 1. Species, cultivar and seed source for cultivars tested in this study.

Species	Cultivar	Seed Source
Capsicum annuum	Ruby King	Seed Savers
Capsicum annuum	Sweet Big Daddy Hybrid	Burpee
Capsicum baccatum	Dedo De Moca	PepperLover.com
Capsicum baccatum	Aji Limon	Chile Pepper Institute
Capsicum chinense	Tobago Yellow Scotch	PepperLover.com
Capsicum chinense	Trin. Scorpion Sweet	PepperLover.com
Capsicum chinense	Hot Zavory	Burpee
Capsicum eximium	CGN 19198	PepperLover.com
Capsicum eximium	CGN 24332	PepperLover.com
Capsicum frutescens	Bradley	PepperLover.com
Capsicum frutescens	Tobasco Heirloom	Burpee
Capsicum frutescens	Tobasco	Seed Savers
Capsicum pubescens	Red Rocoto	Seed Savers

2.2. Cloning Experiments

Once plants formed stems with a diameter greater than 1 mm, branches were cut for cloning experiments and placed into an aeroponic cloning chamber. Cuttings ranging from 7.62–15.24 cm in length were taken from top growing tips or hearty side shoots. Each cutting had 2–4 sets of leaves to aid photosynthetic activity. The lower two nodes and any flower buds were trimmed from cuttings. Sterile clean blades were used to make a 45-degree angle cut 0.635 cm below a set of nodes, and cuttings were placed in stem collars, allowing a minimum of 2.54 cm of the stem to be exposed under the collar. For all aeroponic cloning experiments, as recommended by the manufacturer, 50 mL of Power Clone Advanced Liquid Formula (Botanicare, AZ, USA) was mixed per 3.79 L of distilled water and the solution was placed in the reservoir of a Turbo Klone T96 cloning machine (Everything Green Hydroponics, LLC. Las Vegas, NV, USA) and run with continuous spraying to help increase the onset of node formation (Figure 1). Two Sun Blaze Fluorescent Strip Lights were placed 26.67 cm apart from each other and 30.48 cm above pods in the growing chamber. Plants were grown at 21.0 °C. Each clone was inspected for; (1) time to node formation, identified as time from stem cutting to the initiation of node formation, (2) whole plant dry weight 14-days post node formation, and (3) root dry weight 14-days post node formation.

Figure 1. *Capsicum* spp. cuttings in aeroponic cloning machine.

2.3. Statistical Analysis

The experimental design was a randomized complete block with three blocks, where treatments consisted of different *Capsicum* species. Stem width and height at the time of cloning were treated as covariates. Analyses were conducted in the R statistical software package [20]. Treatment means were separated using a Fisher's protected least significant difference with a significance level of $P < 0.05$. When treatments did not grow, they were treated as missing data.

3. Results and Discussion

There were significant differences in vigor and cloning success between species in time to node formation and the robustness of subsequent cloned material (Figure 2). As expected, *C. annuum* was the fastest species to form nodes. *C. annuum* is the most highly cultivated of the species tested in this study and has the longest domestication history. Unexpectedly, *C. chinense* had the slowest time to node formation, as its growth habit suggested that it would be amenable to cloning, given that stem diameter is a highly significant covariate to cloning success. *Capsicum eximium*, the only non-domestic species, failed to resume active growth after cloning (Figure 2). This is likely due to small stems of *C. eximium* (Figure 3E,F), as well as the wild nature of the plant. Future studies should use larger cuttings with thicker stems to evaluate if aeroponic cloning may be suitable for exotic (non-domestic) species. If this is successful, aeroponic cloning may aid in the preservation of rare species, some of which are endangered and poorly conserved in ex situ collections. One such example is *C. galopengense*, which is no longer present in its native Galapagos Islands due to the introduction of goat herbivory [21].

It took an average of 15.6 days for all species to form a node, similar to what previous studies reported for germination time in pepper [22]. *Capsicum chinense* seeds are known to take longer to germinate and obtain shorter heights than *C. annuum* plants [23]. This trend is similar to the findings from this study for node formation and root emergence. The range of days to growth initiation was much smaller for cloned material in this experiment (10–28 days) as compared to previously reported seed germination ranges of 10–47.4 days [22]. Our data supports previous research showing that somatic regeneration of pepper is feasible [24,25]. Plant growth proceeded normally in all cloned material. Initial stem width and stem height at the time of cloning (Figure 3) were significant covariates ($P < 0.01$). There were initial differences in growth at seven days after node formation for both total dry weight and ratio of above to below ground biomass (Figure 2). However, after 14 days these differences were minimal. Dissimilarities in plant vigor were indicative of ease of cloning and provided an opportunity to assess which species are most amenable to this procedure.

Aeroponic cloning chambers are highly accessible to small-scale growers, requiring only $300 USD to purchase the chamber, nutrient solution, and lighting, as compared to the tissue culture approach that can have operating costs of approximately $24,600 USD per year (numbers based on an estimate for St. Paul, Minnesota, 2016). There was an increased labor requirement relative to traditional crossing, but the cloning methodology provides an opportunity to preserve genotypes that cannot be maintained as seed stocks. Additionally, aeroponic cloning may help small scale boutique growers by allowing them to increase production of novel genotypes without the need for true breeding varieties. In this study the success of this method depended on propagation material being disease and insect free, as plants treated with insecticide or fungicide showed a lack of node formation, root formation, and overall growth. Therefore, cuttings should be taken from plants either grown indoors and disease free, or from outdoor plants pretreated for pests prior to taking cuttings for cloning.

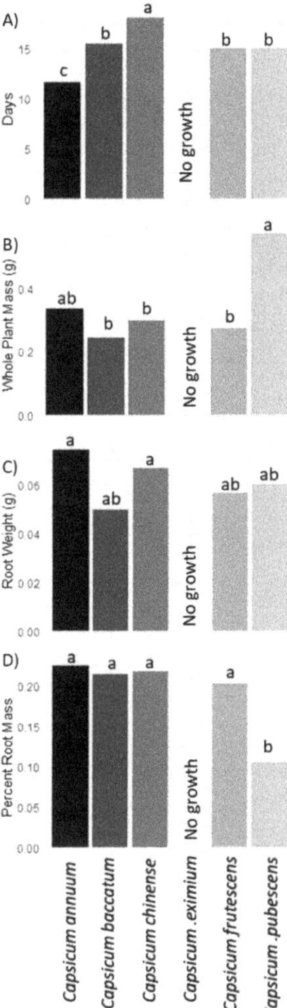

Figure 2. Vigor measurements from different *Capsicum* spp. tested in aeroponic conditions. (**A**) Time to node formation in days, (**B**) Total Dry Weight in grams (g), (**C**) Root Dry Weight in grams (g), (**D**) Ratio of Root Dry weight to Shoot Dry Weight. Letters indicate significant differences between means based on a Fischer's protected LSD.

Figure 3. Variation of stem thickness for *Capsicum* spp. evaluated in this study: *C. annuum* (**A**), *C. baccatum* (**B**), *C. chinense* (**C**), *C. frutescens* (**D**), *C. eximium* strain CGN 19198 (**E**), and *C. eximium* strain CGN 24332 (**F**). Yellow scale bar = 1 cm.

4. Conclusions

The success of cloning in all five domesticated *Capsicum* spp. in aeroponic conditions illustrates that it is a viable option for increasing populations of plants with desirable phenotypic traits for both home and boutique growers. Aeroponic cloning offers a cost-effective solution for increased propagation, as some seeds of highly sought after varieties are expensive and are available only in limited quantities. Future experiments should explore the yield and quality of cloned material in both the field and the greenhouse.

Author Contributions: Conceptualization, D.J.B., T.H.; methodology, D.J.B. analysis, M.B.K., S.M., A.R.D.V.-E.; investigation, J.B., M.K.F., L.H.; resources, D.J.B.; data curation, D.J.B., J.B., M.K.F., L.H.; writing—original draft preparation, M.B.K., S.M., A.R.D.V.-E. writing—review and editing, D.J.B., M.B.K., S.M., A.R.D.V.-E.; visualization, M.B.K., D.J.B., A.R.D.V.-E.

Funding: This research received no external funding.

References

1. Bosland, P.W.; Votava, E.J. *Peppers: Vegetable and Spice Capsicums*; CAB Intl.: Wallingford, UK, 2012.
2. Omolo, M.O.; Wong, Z.Z.; Mergen, A.K.; Hastings, J.C.; Le, N.C.; Reiland, H.A.; Case, K.A.; Baumler, D.J. Antimicrobial properties of chili peppers. *J. Infect. Dis. Ther.* **2014**, *2*, 145. [CrossRef]
3. Sherman, P.W.; Billing, J. Darwinian Gastronomy: Why We Use Spices Spices taste good because they are good for us. *BioScience* **1999**, *49*, 453–463. [CrossRef]
4. Smith, S.H. In the shadow of a pepper-centric historiography: Understanding the global diffusion of capsicums in the sixteenth and seventeenth centuries. *J. Ethnopharmacol.* **2015**, *167*, 64–77. [CrossRef] [PubMed]
5. Kraft, K.H.; Brown, C.H.; Nabhan, G.P.; Luedeling, E.; Ruiz, J.D.J.L.; d'Eeckenbrugge, G.C.; Hijmans, R.J.; Gepts, P. Multiple lines of evidence for the origin of domesticated chili pepper, Capsicum annuum, in Mexico. *Proc. Natl. Acad. Sci. USA* **2014**, *111*, 6165–6170. [CrossRef] [PubMed]

6. Perry, L.; Dickau, R.; Zarrillo, S.; Holst, I.; Pearsall, D.M.; Piperno, D.R.; Berman, M.J.; Cooke, R.G.; Rademaker, K.; Ranere, A.J.; et al. Starch fossils and the domestication and dispersal of chili peppers (*Capsicum* spp. L.) in the Americas. *Science* **2007**, *315*, 986–988. [CrossRef] [PubMed]

7. FAOSTAT. Final Data 2014. 2014. Available online: http://www.fao.org/faostat/en/#home (accessed on 17 January 2017).

8. USDA ERS. 2017. Available online: https://www.nass.usda.gov/Data_and_Statistics/index.php (accessed on 17 February 2017).

9. Bosland, P.W. Capsicums: Innovative uses of an ancient crop. In *Progress in New Crops*; Janick, J., Ed.; ASHS Press: Arlington, VA, USA, 1996; pp. 479–487.

10. Zewdie, Y.; Bosland, P.W. Capsaicinoid profiles are not good chemotaxonomic indicators for Capsicum species. *Biochem. Syst. Ecol.* **2001**, *29*, 161–169. [CrossRef]

11. Lillywhite, J.M.; Simonsen, J.E.; Uchanski, M.E. Spicy Pepper Consumption and Preferences in the United States. *HortTechnology* **2013**, *23*, 868–876. [CrossRef]

12. Bosland, P.W.; Coon, D.; Reeves, G. 'Trinidad Moruga Scorpion' Pepper is the World's Hottest Measured Chile Pepper at More Than Two Million Scoville Heat Units. *HortTechnology* **2012**, *22*, 534–538. [CrossRef]

13. Kantar, M.B.; Anderson, J.E.; Lucht, S.A.; Mercer, K.; Bernau, V.; Case, K.A.; Le, N.C.; Frederiksen, M.K.; DeKeyser, H.C.; Wong, Z.Z.; et al. Vitamin Variation in *Capsicum* Spp. Provides Opportunities to Improve Nutritional Value of Human Diets. *PLoS ONE* **2016**, *11*, e0161464. [CrossRef] [PubMed]

14. Tewksbury, J.J.; Reagan, K.M.; Machnicki, N.J.; Carlo, T.A.; Haak, D.C.; Peñaloza, A.L.C.; Levey, D.J. Evolutionary ecology of pungency in wild chilies. *Proc. Natl. Acad. Sci. USA* **2008**, *105*, 11808–11811. [CrossRef] [PubMed]

15. Kraft, K.H.; de Luna-Ruíz, J.; Gepts, P. Different seed selection and conservation practices for fresh market and dried chile farmers in Aguascalientes, Mexico. *Econ. Bot.* **2010**, *64*, 318–328. [CrossRef] [PubMed]

16. Mehandru, P.; Shekhawat, N.S.; Rai, M.K.; Kataria, V.; Gehlot, H.S. Evaluation of aeroponics for clonal propagation of *Caralluma edulis*, *Leptadenia reticulata* and *Tylophora indica*—Three threatened medicinal Asclepiads. *Physiol. Mol. Biol. Plants* **2014**, *20*, 365–373. [CrossRef] [PubMed]

17. Harlan, J.R.; de Wet, J.M.J.; Price, E.G. Comparative evolution of cereals. *Evolution* **1973**, *27*, 311–325. [CrossRef] [PubMed]

18. Gogoi, S.; Acharjee, S.; Devi, J. In vitro plantlet regeneration of *Capsicum chinense* Jacq. cv.'Bhut jalakia': Hottest chili of northeastern India. *In Vitro Cell. Dev. Biol. Plant* **2014**, *50*, 235–241. [CrossRef]

19. Ritter, E.; Angulo, B.; Riga, P.; Herran, C.; Relloso, J.; San Jose, M. Comparison of hydroponic and aeroponic cultivation systems for the production of potato minitubers. *Potato Res.* **2001**, *44*, 127–135. [CrossRef]

20. R Core Team. *R: A Language and Environment for Statistical Computing*; R Foundation for Statistical Computing: Vienna, Austria, 2015.

21. Tye, A. The status of the endemic flora of Galapagos: The number of threatened species is increasing. *Galápagos Rep.* **2006**, *2007*, 96–103.

22. Hernández-Verdugo, S.; Oyama, K.; Vázquez-Yanes, C. Differentiation in seed germination among populations of *Capsicum annuum* along a latitudinal gradient in Mexico. *Plant Ecol.* **2001**, *155*, 245–257. [CrossRef]

23. Randle, W.M.; Honma, S. Dormancy in peppers. *Sci. Hortic.* **1981**, *14*, 19–25. [CrossRef]

24. Dabauza, M.; Pena, L. High efficiency organogenesis in sweet pepper (*Capsicum annuum* L.) tissues from different seedling explants. *Plant Growth Regul.* **2001**, *33*, 221–229. [CrossRef]

25. Kumar, V.; Gururaj, H.B.; Prasad, B.N.; Giridhar, P.; Ravishankar, G.A. Direct shoot organogenesis on shoot apex from seedling explants of *Capsicum annuum* L. *Sci. Hortic.* **2005**, *106*, 237–246. [CrossRef]

 *horticulturae*

Article

Propagation from Basal Epicormic Meristems Remediates an Aging-Related Disorder in Almond Clones

Thomas Gradziel [1,*], Bruce Lampinen [1] and John E. Preece [2]

1 Department of Plant sciences, University of California, Davis, CA 95616, USA
2 USDA/ARS Germplasm Repository, Davis, CA 95616, USA
* Correspondence: tmgradziel@ucdavis.edu; Tel.: +1-530-752-1575

Received: 5 February 2019; Accepted: 7 April 2019; Published: 11 April 2019

Abstract: The asexual propagation of clonal crops has allowed cultivation of superior selections for thousands of years. With time, some clones deteriorate from genetic and epigenetic changes. Non-infectious bud-failure (NBF) in cultivated almond (*Prunus dulcis*) is a commercially important age-related disorder that results in the failure of new vegetative buds to grow in the spring, with dieback of terminal shoots, witches-brooming of surviving buds, and deformed bark (roughbark). The incidence of NBF increases with clone age, including within individual long-lived trees as well as nursery propagation lineages. It is not associated with any infectious disease agents. Consequently, nursery practices emphasize the establishment of foundation-mother blocks utilizing propagation-wood selected from proven and well-monitored propagation-lineages. Commercial propagation utilizes axillary shoot buds through traditional budding or grafting. This study examines NBF development using basal epicormic buds from individual trees of advanced age as an alternative source of foundation stock. Results show the age-related progression of NBF is suppressed in these epicormic meristems, possibly owing to their unique origins and ontogeny. NBF development in commercial orchards propagated from foundation blocks established from these sources was similarly dramatically suppressed even over the 10- to 20-year expected commercial orchard-life. Foundation-stock stability can be further maintained through appropriate management of propagation source-trees, which requires accurate knowledge of meristem origin and development.

Keywords: clone aging; foundation-stock; genetic-disorder; non-infectious; epigenetic

---

1. Introduction

Vegetative propagation allows the capture and immortalization of horticulturally desirable genetic/genomic configurations with the estimated age of some vegetatively propagated fig and grape clones at over 1000 years [1]. The deterioration of clone performance with age has similarly been reported in several crops [2–4] due to genetic [4] or epigenetic [5,6] changes resulting in losses in productivity and/or crop value. Non-infectious bud-failure (NBF) is a disorder of almond (*Prunus dulcis*, DA Webb) characterized by the failure of terminal vegetative buds to push in the spring [7]. Because new shoot and branch development is necessary in almond to produce new flowers and fruit-bearing wood, NBF development often results in a dramatic loss of crop productivity, particularly when occurring early in orchard tree structural development [8]. Once established, it is irreversible [9] with expression becoming more extensive with increasing time both in individual affected trees as well as in nursery propagation-lineages. Most tree crop nurseries maintain distinct propagation-lineages (i.e., clonal propagation sources that can be traced back through earlier propagations to a common source) to preserve the integrity and quality of their clones. To ensure availability of propagation material that is true-to–type and free from virus-infection and genetic disorders, commercially important clonal propagation sources are maintained in special foundation blocks which are quarantined to control virus contamination and routinely monitored for quality maintenance. Foundation propagation stock from these blocks is then used to establish commercial nursery 'mother-blocks' which, in turn, provide the quality and quantity of buds required for commercial plantings. NBF was first reported in California almond orchards in the early 1930s [10], and has resulted in the commercial decline of many otherwise valuable cultivars and currently threatens the commercial viability of Nonpareil and Carmel, the two most extensively planted California cultivars [8]. Nonpareil, the dominant commercial cultivar, currently at 162,000 ha [11], was originally released in the 1860s with NBF first reported about 1930 [10]. Carmel, the second most important commercial cultivar with current plantings of approximately 53,000 ha was released in 1964 with NBF first being observed in 1978 [12]. Extensive research has failed to find any association of NBF with known pathogens, including viruses and viroids [13–17]. A comprehensive evaluation of nursery foundation stock for the highly NBF vulnerable cultivar Carmel demonstrated lower incidence in orchards propagated from 'relic-trees' (i.e., trees closely related by their propagation lineage to the original Carmel seedling tree) [18]. Consequently, most current nursery production of both Carmel as well as Nonpareil are based on foundation-stock originating from epicormic shoots developing at the base of 50 to 100 year-old 'relic' trees of Carmel and Nonpareil, respectively. Epicormic shoots in almond are defined as neoformed shoots that are produced from latent meristems that have remained dormant until a signal causes them to grow [19]. They differ from the axillary buds used in traditional propagations in their origins and development, having similarities with the better-studied epicormic growth in oaks [20,21].

Epicormic shoots appear to originate as axillary and associated subtending meristems that initially did not grow and became covered with bark [22]. Thus, the ontological developmental phase at the time of the formation is captured. In *Quercus rubra*, when the main stem (trunk) of trees was cut into 40 cm segments and green shoots forced, they rooted differentially from a rate of 70% for the bottom and most basal segment to 17% on segments cut from 3–4 m above ground [23] demonstrating that the oldest epicormic buds can be quite different in their ontogeny and subsequent development than younger buds. The increase in NBF expression potential with increasing tree and clone age may thus be similar to epicormic meristem development in the better-studied forest tree species. This study examines long-term NBF development when using basal epicormic buds from individual trees of advanced age rather than traditional axillary buds as the propagation source.

2. Materials and Methods

Standard commercial propagation sources (source-clones) for the varieties Carmel and Nonpareil were compared against propagation sources from relic-tree basal epicormic origins (REO). Carmel and Nonpareil test trees were planted in alternate rows with pollinizer trees to provide cross-pollination. All trees were propagated on Nemaguard rootstock using traditional T-budding of axillary buds from previous season's growth of propagation source trees. Source-clone mother trees were either from traditional nursery propagation lineages or one to two generations removed from a REO source (Tables 1 and 2). For Carmel, at least 180 trees per source-clone as well as 31 trees propagated from the original Carmel seedling tree were evaluated. For Nonpareil, 60 trees of each source-clone were tested. Planting took place in 1990 near Wasco, California in the southern San Joaquin Valley where high summer temperatures have been shown to be conducive to early NBF development [18]. Planting was carried out in randomized groups of 10 trees each, in north-south rows with a 7.3 × 6.4 m spacing on a Wasco-Sandy-loam soil. Standard orchard management practices were applied to all treatments.

Table 1. Carmel almond propagation sources.

Source	FPS ID [1]	Origin
E1	na	Delta-mother tree propagated from original Carmel seedling tree
E2	3-56-1-90	Delta source tree 2 propagated basal epicormic bud from Delta tree
E3	3-56-8-92	Delta source tree 13 propagated basal epicormic bud from Delta tree
E4	3-56-2-90	Delta source tree 7 propagated basal epicormic bud from Delta tree
C1	na	Standard commercial nursery source established in 1988
C2	na	Standard commercial nursery source established in 1986
C3	na	Standard commercial nursery source established in 1985
C4	na	Standard commercial nursery source established in 1981

[1] last 2 digits of Foundation Plant Services identification number (FPS ID) denote year of establishment.

Table 2. Nonpareil almond propagation sources.

Source	FPS ID [1]	Origin
E1	3-8-6-72	Propagated from a basal epicormic bud of a 108-year-old relic tree
E2	3-8-17-92	Propagated from a basal epicormic bud of a 100+ year-old relic tree
E3	3-8-2-70	Propagated from a basal epicormic bud of a 100+ year-old heat-treated relic tree
E4	na	Commercial nursery propagation source originating from 100+ year-old relic tree
C1	na	Standard commercial nursery source established in 1988
C2	na	Standard commercial nursery source established in 1986
C3	na	Standard commercial nursery source established in 1984
C4	na	Standard commercial nursery source established in 1977

[1] last 2 digits of Foundation Plant Services identification number (FPS ID) denote year of establishment.

Four Carmel REO source-clones were evaluated. One tree (subsequently designated as the Delta mother tree) was propagated from basal buds of the original Carmel seedling tree with 3 trees subsequently propagated from this Delta mother tree (Foundation Plant Services (FPS) sources 3-56-1-90, 3-56-8-92 and 3-56-2-90) (Table 2). Four standard commercial propagation sources (C1–C4) of Carmel were provided by cooperating nurseries as part of a larger evaluation of NBF risk in commercially available California propagation stocks [18].

Four REO source-clones for Nonpareil were derived from 100+ year old trees from orchards in the cooler northern Sacramento Valley (Table 1). Adventitious epicormic shoots either spontaneously pushing from intact trees or artificially pushed by partial tree girdling/scaffold removal (Figure 1) were used to establish three source-clones (3-8-6-72, 3-8-17-92 and 3-8-2-70) at the University of California FPS Foundation Orchard [24]. An additional REO source designated (E3) was developed by a cooperating commercial nursery. Four standard commercial propagation sources (C1–C4) of Nonpareil were provided by cooperating nurseries as part of a larger evaluation of NBF risk in commercially available California propagation stock [18].

Figure 1. Low non-infectious bud-failure (NBF) foundation stock is developed from epicormic buds (**b,d**) pushed from relic trees (planted near the time of cultivar origin) (**a**). Resulting foundation stock (**c**) is maintained by aggressive annual pruning to promote both axillary and epicormic shoot growth from older basal wood. (Circled top branch in (**a**) shows classic NBF symptoms of terminal shoot suppression and resulting erratic branching patterns).

Trees were evaluated using established ratings for NBF expression [18] over the next 18 years or until plot removal due to excessive tree decline and loss of commercial productivity. Trees were evaluated in March following primary shoot development. Initial NBF incidence was identified as a failure of terminal shoot bud growth resulting in the induction and growth of sub-apical buds resulting in a distinct terminal shoot die-back as shown in 'a' of Figure 2. Once identified, NBF was verified in subsequent years by a consecutive pattern of terminal shoot failures (as in 'b', 'c', and 'd' in Figure 2) resulting in the distinct 'crazy-top' growth pattern characteristic of NBF.

Figure 2. Shoot expression of NBF. Terminal vegetative buds fail to grow shifting growth to a sub-apical bud (a). The pattern is repeated here with three previous seasons' shoot growth-failures (b, c, and d, respectively) present, resulting in a distinct pattern known as 'crazy-top'. Note that flower buds are not directly affected.

3. Results

3.1. Carmel

All FPS REO clone-sources showed no occurrence of NBF until year 8 of the study (Table 3). In contrast, all commercial sources showed NBF by year 4 with commercial clone-source C4 showing almost 100% of trees affected at this time. The clone-source C4 was also the oldest traditional commercial source-clone tested, having been established in 1981. FPS REO clone-source E4 went from 0% to 51% incidence in year 7, rapidly increasing to almost 89% by year 10, the last year before orchard removal due to loss of commercial productivity (Figure 3). The remaining FPS REO clone-sources E1, E2 and E3 showed NBF expression of only 3%, 5% and 11%, respectively, at year 8, with a gradual increase in incidence to 13%, 8% and 15%, respectively, by year 10.

Table 3. Development of NBF over a 10 year of evaluation period for Carmel relic-tree basal epicormic origins (REO) source-clAones E1–E4 compared with standard commercial sources C1–C4.

	E1		E2		E3		E4		C1		C2		C3		C4	
Year	Bf Trees	Per-cent	Bf Trees	Per-cent	Bf Trees	Per-cent	Bf Trees	Per-cent	Bf Trees	Per-cent	Bf Trees	Per-cent	Bf Trees	Per-cent	Bf Trees	Per-cent
1	0	0	0	0	0	0	0	0	0	0	0	0	0	0	0	0
2	0	0	0	0	0	0	0	0	0	0	0	0	31	10.7	68	36.0
3	0	0	0	0	0	0	0	0	0	0	0	0	36	12.4	147	77.8
4	0	0	0	0	0	0	0	0	60	14.7	53	22.5	102	35.1	183	96.8
5	0	0	0	0	0	0	0	0	71	17.4	66	28.0	114	39.2	189	100
6	0	0	0	0	0	0	0	0	114	28.0	102	43.2	156	53.6	189	100
7	0	0	0	0	0	0	152	51.2	145	35.6	181	76.7	199	68.4	189	100
8	1	3.2	11	5.1	25	10.7	221	74.4	166	40.8	218	92.4	235	80.8	189	100
9	2	6.5	15	6.9	31	13.2	238	80.1	197	48.4	229	97.0	286	98.3	189	100
10	4	12.9	17	7.8	34	14.5	264	88.9	213	52.3	236	100	291	100	189	100
Total	31		217		234		297		407		236		291		189	

Figure 3. NBF expression in a nine-year-old tree where NBF was first observed in year 5, resulting in abnormal tree growth and loss of commercial productivity.

3.2. Nonpareil

Complete control of NBF was achieved in all FPS REO clone sources over the 18-year trial (Table 4). In the commercially developed REO source E4, 5% of trees showed NBF symptoms by year 18, the final year before orchard removal. All commercial sources showed NBF by year 10, with incidents ranging from 2% to 22%. Final, year 18, incidence of NBF in commercial sources ranged from 35% to 87%. Commercial clone-source C4, which showed the highest NBF incidents and rate of increase, was also the oldest traditional commercial source-clone tested, having been established in 1977.

Table 4. Development of NBF over an 18-year evaluation period for Nonpareil REO source-clones E1–E4 compared with standard commercial sources C1–C4. (Sixty trees of each clone-source were evaluated).

Year	E1 BF Trees	E1 Per-cent	E2 BF Trees	E2 Per-cent	E3 BF Trees	E3 Per-cent	E4 BF Trees	E4 Per-cent	C1 BF Trees	C1 Per-cent	C2 BF Trees	C2 Per-cent	C3 BF Trees	C3 Per-cent	C4 BF Trees	C4 Per-cent
1	0	0	0	0	0	0	0	0	0	0	0	0	0	0	0	0
2	0	0	0	0	0	0	0	0	0	0	0	0	0	0	0	0
3	0	0	0	0	0	0	0	0	0	0	0	0	0	0	0	0
4	0	0	0	0	0	0	0	0	0	0	0	0	0	0	0	0
5	0	0	0	0	0	0	0	0	0	0	0	0	0	0	0	0
6	0	0	0	0	0	0	0	0	0	0	0	0	0	0	1	1.7
7	0	0	0	0	0	0	0	0	0	0	0	0	0	0	2	3.3
8	0	0	0	0	0	0	0	0	0	0	0	0	1	1.7	3	5.0
9	0	0	0	0	0	0	0	0	1	1.7	1	1.7	2	3.3	7	11.7
10	0	0	0	0	0	0	0	0	1	1.7	2	3.3	3	5.0	13	21.7
11	0	0	0	0	0	0	0	0	2	3.3	4	6.7	6	10.0	22	36.7
12	0	0	0	0	0	0	0	0	3	5.0	7	11.7	10	16.7	28	46.7
13	0	0	0	0	0	0	0	0	4	6.7	10	16.7	14	23.3	40	66.7
14	0	0	0	0	0	0	0	0	9	15.0	15	25.0	17	28.3	44	73.3
15	0	0	0	0	0	0	0	0	11	18.3	18	30.0	20	33.3	48	80.0
16	0	0	0	0	0	0	0	0	14	23.3	21	35.0	26	43.3	50	83.3
17	0	0	0	0	0	0	0	0	20	33.3	23	38.3	28	46.7	52	86.7
18	0	0	0	0	0	0	3	5	21	35.0	25	41.7	34	56.7	53	88.3

4. Discussion

In both studies, commercial trees propagated from REO sources resulted in a dramatic reduction in overall incidence and delayed initial appearance of NBF when compared to traditional commercial nursery sources (Figures 2 and 3). In every study, once NBF was identified in a population, its subsequent incidence always increased. As populations approached 100% affected, the rate of increase would understandably decrease, although no declines were observed in the proportion of affected trees. Greater control was observed for Nonpareil clonal sources than for Carmel when using either REO or traditional commercial sources.

Establishing REO-based foundation-stock is effective in remediating NBF clonal degeneration in almond clones. Clone quality deterioration with age is an inescapable consequence of the immortalization of tree crop genotypes by consecutive, long-term vegetative propagations [2–5]. The mechanisms remain unknown and appear complex, obviating the otherwise promising opportunities for DNA sequence analysis [5,25]. Vegetative progeny testing, while tedious and time-consuming, remains the only reliable strategy for assessing and maintaining clone integrity. For this reason, many commercial tree crop nurseries have depended upon highly-monitored clonal lineages to maintain the integrity of their particular clone-source. This type of sequential propagation from a common well-characterized source also allows intra-clonal selections for traits such as improved productivity that commercial nurseries exploit to distinguish and market their particular clone-sources. Based on this paradigm of maintenance/improvement through intra-clonal sequential selections within unique nursery clone-sources, early nursery efforts to control NBF emphasized selecting individual trees from such established, elite clonal-lineages that were free from symptoms in otherwise highly symptomatic orchard plantings, for use as future 'improved' propagation sources. Nursery propagation lineages were thus analogous to tree growth where sequential tree-branch ramifications represent sequential generations of propagation-wood selection from more recently established orchards (which possessed the shoot vigor required for the large amount of budwood needed commercially).

NBF has been shown to be a disorder that increases both in individual trees as they age [9,14,26] as well as individual propagation lineages [12,17,18]. Consequently, just as the lowest NBF incidence and so best propagation source would be expected at the base of the tree, the lowest probability of NBF would be expected at the base of the clonal-lineage, which would be the original seedling tree from which the cultivar was initially selected. The variety Nonpareil resulted from a seedling first grown in the 1860s [10,27] with NBF not being reported until the 1930s despite being extensively planted. This indicates that the original seedling tree had a very low probability of NBF that gradually increased with further tree growth as well as increasing generations of propagation from the original seedling-source. Fortunately, by 1960 when Kester [28], initially recognized the relation of NBF with clone aging and, therefore, the potential remediation through propagation from earlier clonal-sources, many northern Sacramento Valley orchards still contained individual Nonpareil trees 100+ years old that were propagated near the time of initial cultivar release. The epicormic buds, ontogenetically, had lower potential for NBF expression than subsequent buds that formed on resulting propagules. Therefore, it is possible that propagation from these buds could restore the original 'normal' phenotype, at least for several years to decades. Returning to these old relic trees for collecting propagation wood thus represents a successful strategy for avoiding NBF.

More than 400,000 ha are currently planted to almonds in California, with over 162,000 ha at 100 to 120 trees per acre, planted to Nonpareil as the dominant variety [11]. (Almond is self-sterile, requiring both early and late season pollenizers to achieve commercial yields). The propagation of the 40,000,000 Nonpareil trees currently in commercial production would have required extensive sequential propagation from multiple propagation lineages, resulting in the variability observed in Table 4. By the time this test orchard was planted in 1990, most commercial nurseries had started to recognize the value of REO propagation sources and begun to shift to these sources, although many were understandably reluctant to abandon their well-established, well-characterized and well-branded propagation lineages. While, Nonpareil clone-source C4, represents a common commercial source prior to 1990, the other commercial sources represent such initial nurseries transition efforts. Based on this and smaller NBF test plots, most modern commercial nursery propagation of Nonpareil is based on the REO FPS sources evaluated in this study (Table 1). The eventual breakdown of Nonpareil clonal-source C4, a nursery selection from a separate but similar REO source, may indicate a vulnerability unique to that source, or alternatively, that all REO sources may be eroding over the subsequent 40–50 years from the original selection, similar to the erosion that occurred from the original seedling tree until first expression in the 1930s.

Commercial propagation of the Carmel variety from REO sources delayed and diminished NBF expression, although because this is a progressive disorder, all FPS REO clonal-sources showed NBF by year 10. Previous, related studies have shown that this cultivar can still be commercially profitable if NBF is delayed until after development of the key scaffolds and branches making up the basic tree bearing-architecture, which is usually about year 7 [18]. The reduced propagation options for NBF control for Carmel parallel the reduced time from cultivar introduction in 1964 to the first reported incidents of NBF in 1977 [18]. While propagation from the Delta mother tree or even the original Carmel seedling tree should result in further reductions in NBF progeny trees, the number of buds available from these individual trees inherently limits the number of trees possible. At its current acreage of 13% of the currently planted 400,000 ha almond acres, or 53,000 ha at 100 to 120 trees per acre, the orchard replenishment would require large nursery propagation mother blocks and so inherently multiple propagation cycles from the initial seedling tree. Consequently, while individual REO-based low-NBF propagation sources are available for this cultivar, commercial scale-up is problematic and, as a result, this cultivar has declined from being the second most extensively planted variety in 1990 at almost 30% of the acreage to only 13% of the acreage at present [11].

While epicormic meristems have been shown to be valuable sources for the remediation of NBF in almond, the mechanism remains unknown. Epicormic meristems are unique in that they can be very long-lived yet exist at only rudimentary stages of development [20,21]. Their morphology and ontogeny also allows them to be arrested at a specific point in development. In contrast, recently-formed axillary meristems normally used in almond propagation are histogenetically complex and need to be regularly renewed.

Basal axillary buds could be the result of recent epicormic shoot initiation and development or could be the consequence of multiple growth-cycles in the often inherently slow-growing lower canopy shoots. Thus, differences between older epicormic and recently-formed axillary bud ontogeny is important because the extensive tree hedging currently recommended to suppress clone aging in FPS foundation blocks [14] (Figure 1) tends to push both epicormic and axillary buds for the next generation of propagation budwood collection.

5. Conclusions

This study examined NBF development using basal epicormic buds from individual trees of advanced age (REO sources) as an alternative source of propagation stock. Results show the age-related progression of NBF is suppressed in these epicormic meristems, possibly owing to their unique origins and ontogeny. Long-term NBF development in commercial orchards propagated from foundation blocks established from these sources was also dramatically suppressed, even over the 10- to 20-year expected commercial orchard-lifetimes. Thus, REO-based foundation stock was effective in suppressing NBF, an age-related disorder in almond. The extent and so commercial value of this suppression varied with cultivar and was related to NBF expression potential of the initial seedling tree for each cultivar and the time or number of generations of sequential vegetative propagations until NBF first appeared. Because the nature and ontogeny of epicormic meristems are largely uncharacterized for this as well as most tree crops, the underlying mode of action for suppression remains unknown and vegetative-progeny testing remains the only effective strategy for identifying suitable commercial propagation clone-sources.

Author Contributions: Conceptualization, T.G. and B.L.; Methodology, T.G.; Validation, T.G., B.L. and J.E.P.; Formal Analysis, T.G., B.L., J.E.P.; Investigation, T.G., B.L., J.E.P.; Resources, T.G.; Writing—Original Draft Preparation, T.G.; Writing—Review and Editing, T.G., J.E.P., B.L.; Visualization, T.G., J.E.P.; Supervision, T.G.; Project Administration, T.G.; Funding Acquisition, T.G., B.L.

Funding: This research was funded by Almond Board of California and the California Nurseries Advisory Board Funding Number 498B with support from the University of California at Davis.

Acknowledgments: The authors wish to acknowledge support given by Paramount Orchards, Wasco California and Billings Ranch, Delano California in providing fields and land management for these valuations.

Conflicts of Interest: The authors declare no conflict of interest.

References

1. Janick, J. The Origins of Fruits, Fruit Growing, and Fruit Breeding. *Plant Breed. Rev.* **2005**, *25*, 255–320.
2. Kester, D.E. The clone in Horticulture. *HortScience* **1983**, *18*, 831–837.
3. Skowcroft, W.R. Somaclonal Variation, the Myth of Clonal Uniformity. In *Genetic Flux in Plants*; Hohn, B., Dennis, E.S., Eds.; Springer: New York, NY, USA, 1985; pp. 217–245.
4. Skirvin, R.M.; McPheeters, K.D.; Norton, M. Sources and frequency of somaclonal variation. *HortScience* **1994**, *29*, 1232–1237. [CrossRef]
5. D'Aquila, P.; Rose, G.; Bellizzi, D.; Passarino, G. Epigenetics and aging. *Maturitas* **2013**, *74*, 130–136. [CrossRef]
6. Fraga, M.F.; Rodriguez, R.; Canal, M.J. Genomic DNA methylation-demethylation during aging and reinvigoration of Pinus radiata. *Tree Physiol.* **2002**, *22*, 813–816. [CrossRef] [PubMed]
7. Kester Dale, E.; Asay, R.N. Variability in noninfectious bud-failure of 'Nonpareil' almond. Location and environment. *J. Am. Soc. Hortic. Sci.* **1978**, *103*, 377–382.
8. Kester, D.E.; Jones, R.W. Noninfectious bud-failure from breeding programs of almond (*Prunus amygdalus* Batsch). *J. Am Soc. Hortic. Sci.* **1970**, *74*, 214–219.
9. Fenton, C.A.L.; Kester, D.E.; Kuniyuki, A.H. A model for the development of noninfectious bud failure in vegetatively propagated populations of almond cultivars. *Hortscience* **1986**, *21*, 738.

10. Wilson, E.E.; Schein, R.D. The nature and development of noninfectious bud-failure in almond. *Hilgardia* **1956**, *24*, 519–542. [CrossRef]

11. Almond Board of California. 2018. Available online: http://www.almonds.com/sites/default/files/Almanac%202018.pdf (accessed on 6 January 2019).

12. Kester, D.E.; Shackel, K.A.; Micke, W.C.; Gradziel, T.M.; Viveros, M. Variability in potential and expression of noninfectious bud-failure among nursery propagules of 'Carmel' almond. *Acta Hortic.* **1998**, *470*, 268–272. [CrossRef]

13. Fenton, C.A.L.; Kuniyuki, A.H.; Kester, D.E. Search for a viroid etiology for noninfectious bud failure in almond. *HortScience* **1988**, *23*, 1050–1053.

14. Kester, D.E.; Gradziel, T.M. Genetic Disorders. In *Almond Production Manual*; Micke, W.C., Ed.; University of California: Oakland, CA, USA, 1996; pp. 76–87.

15. Kester, D.E. Noninfectious Bud-Failure in Almond. In *Virus Diseases and Disorders of Stone Fruits in North America*; Fulton, R.W., Ed.; Agricultural Research Service, U.S. Department of Agriculture: Washington, DC, USA, 1976; pp. 278–283.

16. Kester, D.E. Noninfectious bud-failure, a nontransmissable inherited disorder in almond. I. Pattern of phenotypic inheritance. *Proc. Am. Soc. Hortic. Sci.* **1968**, *92*, 7–15.

17. Kester, D.E. Noninfectious bud-failure, a nontransmissable inherited disorder in almond II. Progeny tests for bud-failure. *Proc. Am. Soc. Hortic. Sci.* **1968**, *92*, 16–28.

18. Kester, D.E.; Shackel, K.A.; Micke, W.C.; Viveros, M.; Gradziel, T.M. Noninfectious bud failure in 'Carmel' almond, I. Pattern of development in vegetative progeny trees. *J. Am. Soc. Hortic. Sci.* **2004**, *129*, 244–249. [CrossRef]

19. Negrón, C.; Contador, L.; Lampinen, B.D.; Metcalf, S.G.; Guédon, Y.; Costes, E.; DeJong, T.M. Differences in proleptic and epicormic shoot structures in relation to water deficit and growth rate in almond trees (*Prunus dulcis*). *Ann Bot.* **2014**, *113*, 545–554. [CrossRef]

20. Fontaine, F.; Kiefer, E.; Clemen, T.C.; Burrus, M.; Druelle, J.L. Ontogeny of the proventitious epicormic buds in Quercus petraea. II. From 6 to 40 years of the tree's life. *Trees Struct. Funct.* **1999**, *14*, 83–90.

21. Meier, A.R.; Saunders, M.R.; Michler, C.H. Epicormic buds in trees: A review of bud establishment, development and dormancy release. *Tree Physiol.* **2012**, *32*, 565–584. [CrossRef]

22. Preece, J.E. Stock Plant Physiological Factors Affecting Growth and Morphogenesis. In *Plant Propagation by Tissue Culture*, 3rd ed.; George, E.F., Hall, M.A., de Klerk, G.J., Eds.; Springer: Dordrecht, The Netherlands, 2008; Volume 1, p. 507.

23. Fishel, D.W.; Zaczek, J.J.; Preece, J.E. Positional influence on rooting of shoots forced from the main bole in swamp white oak and northern red oak. *Can. J. For. Res.* **2003**, *33*, 705–711. [CrossRef]

24. Foundation Plant Services. 2019. Available online: http://fps.ucdavis.edu/index.cfm (accessed on 6 January 2019).

25. Fresnedo-Ramírez, J.; Chan, H.M.; Parfitt, D.E.; Crisosto, C.H.; Gradziel, T.M. Genome-wide DNA-(de)methylation is associated with Noninfectious Bud-failure exhibition in Almond (*Prunus dulcis* [Mill.] D.A.Webb). *Sci. Rep.* **2017**, *7*. [CrossRef]

26. Kester, D.E. Noninfectious bud-failure, a nontransmissable inherited disorder in almonds, III. Variability in BF potential within plants. *J. Am. Soc. Hortic. Sci.* **1970**, *95*, 162–165.

27. Gradziel, T.M. Almond origin and domestication. In *Horticultural Reviews*; Janick, J., Ed.; John Wiley & Sons: Hoboken, NJ, USA, 2011; pp. 23–82.

28. Kester, D.E. Inheritance of noninfectious bud-failure in almond. *Proc. Am. Soc. Hortic. Sci.* **1961**, *77*, 278–285.

 *horticulturae*

Article

In Vitro Establishment of 'Delite' Rabbiteye Blueberry Microshoots

Carolina Smanhotto Schuchovski and Luiz Antonio Biasi *

Departamento de Fitotecnia e Fitossanitarismo, Universidade Federal do Paraná, Rua dos Funcionários, 1540.
CEP 80.035-050, Curitiba, Brazil; carolina.sschu@gmail.com

* Correspondence: biasi@ufpr.br; Tel.: +55-41-3350-5682

Received: 18 January 2019; Accepted: 25 February 2019; Published: 8 March 2019

Abstract: Micropropagation is an important technique for clonal mass propagation and a tool for in vitro studies. One of the first steps to overcome in this process is the establishment of new explants in vitro. 'Delite' rabbiteye blueberry was cultured in vitro with four cytokinins (zeatin (ZEA), 6-(γ-γ-dimethylallylamino)-purine (2iP), 6-benzylaminopurine (BAP), and kinetin (KIN)) at eight concentrations (0, 2.5, 5, 10, 20, 30, 40, and 50 μM). Additionally, nine combinations of nitrogen salts were tested, using Woody Plant Medium (WPM) and a modified WPM as the basic medium. ZEA and 2iP showed better responses, but ZEA was superior at lower (2.5 μM) concentrations (89.7% survival, 81.3% shoot formation, 1.3 shoots, 13.8 mm shoot length, 10.0 leaves). BAP and KIN showed very low responses. In the combinations of salts with modified WPM, no differences were observed. However, the original WPM with treatments of $0.5 \times NH_4NO_3$ and $1 \times Ca(NO_3)_2$, $0.5 \times NH_4NO_3$ and $0.5 \times Ca(NO_3)_2$, and the modified WPM alone showed the lowest rates of survival and shoot formation and the shortest shoot lengths. The highest shoot lengths were observed in treatments with the original WPM, $1.5 \times NH_4NO_3$ and $0.5 \times Ca(NO_3)_2$, and $1.5 \times NH_4NO_3$ and $1.5 \times Ca(NO_3)_2$. This initial study with 'Delite' can be the basis for further experiments with different combinations of salts, 2iP, and ZEA.

Keywords: *Ericaceae; Vaccinium virgatum;* micropropagation; in vitro culture; cytokinins; zeatin; 2iP; BAP; kinetin; WPM

1. Introduction

Blueberry is a woody perennial species in the family Ericaceae and genus *Vaccinium*. The fruit is a true berry with many seeds, with color ranging from light blue to black and a waxy cuticle layer [1]. Blueberry has been gaining great importance in fruit production, especially because of its recognized taste properties and its nutraceutical qualities as an anti-inflammatory and anti-oxidant, being a health-promoting food [2]. Blueberry fruits are rich in polyphenols [3]. These blueberry polyphenols show anti-inflammation activity, related to the balances in pro-inflammatory cytokines, and they could be used as an anti-inflammatory medicine [4]. Among the phenolic compounds that appear at high levels in blueberries are anthocyanins [5], flavonols, and phenolic acids [6]. The anthocyanin found in high amounts in blueberries contributes to preventing several chronic diseases, such as neurodegenerative diseases, cardiovascular disorders, cancer, and diabetes [7].

Much research has been developed related to the propagation of blueberries. Traditionally, blueberry is propagated by softwood, semi-hardwood, and hardwood cuttings [8] or even rhizome cuttings of selected clones [6]. Some challenges in this production are a very low rooting percentage in many genotypes, the amount of time required to propagate and commercialize newly-released cultivars for mass propagation [8,9], and phytosanitary problems. In vitro culture (micropropagation) can overcome the limitations of traditional cuttings, presenting an alternative for faster growth [10] throughout the year (with no seasonal effects) without pathogens [11]. There are some studies on the in vitro propagation of *Vaccinium* species, but only some of these have been done with rabbiteye (*V. virgatum* Ait. (syn. *V. ashei* Reade)), specifically for the 'Delite' rabbiteye cultivar that is suitable for and adapted to regions of southern Brazil. For this specific cultivar, some research concerning in vitro protocol is still required to give more information on the optimal conditions for the development of this technique.

One crucial point in tissue culture techniques is the appropriate use, type, and concentration of growth regulators and the combination of culture medium salts that allows fast, efficient development of the initial explants. Understanding the interference of factors can lead to the development of further regeneration protocols that could be useful for either micropropagation or developing regeneration techniques necessary for plant recovery after cell transformation. There is some research showing that the lack of new shoot growth can make initiation the limiting step in establishing *Vaccinium* cultures in vitro [12]. Studies also show that new growth in vitro is difficult to achieve in *Vaccinium*, especially when using plant material from the field [13].

For the initial phase of in vitro culture, a combination of cytokinins can usually be used. In the initial in vitro culture in one study using nodal segments from softwood cuttings of 'Ozarkblue' highbush blueberry (*V. corymbosum*), zeatin (ZEA) and 6-(γ-γ-dimethylallylamino)-purine (2Ip) were tested in the initial culture medium in different combinations (18 μM ZEA, 25 μM 2-iP, and 9.1 μM ZEA combined with 25 μM 2iP) using WPM as the basal medium. On medium with ZEA present, shoots developed with green and red leaves. However, on medium containing only 2iP, shoots had light red leaves and callus at the base with stunted growth [9].

In lowbush blueberry (*V. angustifolium* Ait.) cultivated in the initiation phase medium containing 5 μM ZEA or 10 μM 2iP, explants produced elongated shoots with both growth regulators. However, ZEA treatments showed a higher percentage of new shoot growth compared to 2iP for three cultivars [6].

Wild bilberry (*V. myrtillus* L.) and lingonberry (*V. vitis-idaea* L.) were tested using buds and shoot tips on a modified MS medium supplemented with 2iP variations from 9.8 to 78.4 μM. For bilberry and lingonberry, the best results were obtained with 49.2 μM and 24.6 μM, respectively. Brownish explants were observed with an increasing 2iP concentration [13]. For 'Berkeley', 'Bluecrop' and 'Earliblue', highbush blueberries, and 'O'Neal' southern highbush blueberry, a medium containing 20 μM ZEA was used in the initiation of cultures [14].

Concerning the type of basal culture medium, many studies have used WPM as the basic medium for blueberry [14]. However, some authors have tried to optimize this medium by making some modifications, such as combining MS and WPM media, creating an MW medium [14], or proposing some changes in the components [15], leading to a modified WPM. A well-balanced medium is important to prevent stunted growth and physiological disorders [16]. Some authors have discussed the importance of the balance between nitrogen forms used in tissue culture (NO_3^- and NH_4^+) as much as the total amount of nitrogen in the culture medium [17].

The objective of this work was to determine an efficient growth regulator and balance of nitrogen salts for the establishment of 'Delite' microcuttings in in vitro culture.

2. Materials and Methods

In this work, three experiments in initial in vitro culture were designed. In the first one, four different cytokinins (ZEA, 2iP, 6-benzylaminopurine (BAP), and kinetin (KIN)) were tested at eight different concentrations. The second experiment tested nine different combinations of the nitrogen salts ($(NH_4)_2SO_4$, KNO_3, and $Ca(NO_3)_2 \cdot 4H_2O$), using the modified WPM [15] as the basic medium. The third experiment tested nine different combinations of two nitrogen salts, NH_4NO_3 and $Ca(NO_3)_2 \cdot 4H_2O$, using the original WPM [18] as the basic medium and compared them with treatment 10 (modified Woody Plant Medium [15]).

2.1. Plant Material

One-year-old hardwood cuttings were collected during winter from field-grown rabbiteye blueberry 'Delite' mother plants at the Experimental Station of Universidade Federal do Paraná, Pinhais/PR. They were treated with an immersion in fungicide solution for 5 min (Cercobin®0.2%) and stored at a 4 °C temperature at the Micropropagation Laboratory, UFPR, Curitiba/PR for one to two months in plastic bags. Cuttings were placed in glass containers with water in the culture room at 25 °C $\pm$ 2 °C under cool day light at 40 μmol m^{-2} s^{-1} with a 16-h photoperiod. Newly formed shoots were collected and used as explants for the establishment of cultures.

Two-node segments (0.8–2 cm in length, discarding the apical portion of the donor-explant) were collected and surface sterilized with 70% (*v/v*) ethanol for 30 s, followed by immersion in 0.5% sodium hypochlorite solution containing 0.1% (*v/v*) Tween 20 for 5 min. They were washed with sterile deionized water three times inside the laminar flow chamber.

2.2. Culture Medium and Growing Conditions

Explants were isolated in culture tubes (150 × 30 mm), with each containing 6 mL of modified culture medium, differing in each of the three experiments. In all experiments, the medium was supplemented with Murashige and Skoog (MS) [19] vitamins, 30 g L^{-1} sucrose, 0.1 g L^{-1} myo-inositol, and 6 g L^{-1} agar (Vetec®). The pH of all media was adjusted to 5.2 before autoclaving at 120 °C and 1.5 atm.

2.2.1. Experiment 1: Cytokinins

Microcuttings were isolated in the modified WPM [2] (Table 1), supplemented as described above. Eight different concentrations (0, 2.5, 5, 10, 20, 30, 40, and 50 μM) of each of the four cytokinin growth regulators, ZEA, 2iP, BAP, and KIN, were tested, for a total of 32 treatments. ZEA and 2iP, when used, were sterilized through 0.22 μm filters and added to the cooled media. BAP and KIN were added to media before autoclaving.

Table 1. Modified Woody Plant Medium (modified WPM) [15] and original WPM [18] culture medium compositions.

Components	Modified WPM	Original WPM
Macronutrients	Final Concentration in the Culture Medium (mg L^{-1})	
$(NH_4)_2SO_4$	119.00	-
NH_4NO_3	-	400.00
KNO_3	893.00	-
K_2SO_4	-	990.00
KH_2PO_4	170.00	170.00
$Ca(NO_3)_2 \cdot 4H_2O$	278.00	556.00
$CaCl_2 \cdot 2H_2O$	-	96.00
$MgSO_4 \cdot 7H_2O$	370.00	370.00
Micronutrients	-	-
$FeSO_4 \cdot 7H_2O$	55.60	27.80
Na_2-EDTA	74.60	37.30
H_3BO_3	6.20	6.20
$MnSO_4 \cdot H_2O$	22.30	22.30
$ZnSO_4 \cdot 7H_2O$	8.60	8.60
KI	0.415	-
$Na_2MoO_4 \cdot 2H_2O$	0.25	0.25
$CuSO_4 \cdot 5H_2O$	0.025	0.25

2.2.2. Experiment 2: Combinations of $(NH_4)_2SO_4$, KNO_3, and $Ca(NO_3)_2 \cdot 4H_2O$ Using the Modified WPM [15] as the Basic Medium

Explants were isolated using nine different treatments as described in Table 2, with different amounts (1×, 0.5× or 1.5×) of $(NH_4)_2SO_4$, KNO_3, and $Ca(NO_3)_2 \cdot 4H_2O$ (Table 2), using the modified

WPM [2] (Table 1) as the basic medium. Media were supplemented as described above, with the addition of cytokinin ZEA (5 µM).

Table 2. Experiment 2 with treatments 1 to 9 on the modified Woody Plant Medium (WPM) [15] with different amounts of $(NH_4)_2SO_4$ (x), KNO_3, and $Ca(NO_3)_2 \cdot 4H_2O$ (x).

Treatments	1 (Modified WPM)	2	3	4	5	6	7	8	9
NH_4: $(NH_4)_2SO_4$ (x)	1×	1×	1×	0.5×	0.5×	0.5×	1.5×	1.5×	1.5×
NO_3: KNO_3 and $Ca(NO_3)_2 \cdot 4H_2O$ (x)	1×	0.5×	1.5×	1×	0.5×	1.5×	1×	0.5×	1.5×
Components	**Final Concentration in the Culture Medium (mg L^{-1})**								
$(NH_4)_2SO_4$	119.0	119.0	119.0	59.5	59.5	59.5	178.5	178.5	178.5
KNO_3	893.0	446.5	1339.5	893.0	446.5	1339.5	893.0	446.5	1339.5
$Ca(NO_3)_2 \cdot 4H_2O$	278.0	139.0	417.0	278.0	139.0	417.0	278.0	139.0	417.0

2.2.3. Experiment 3: Combinations of NH_4NO_3 and $Ca(NO_3)_2 \cdot 4H_2O$ Using the Original WPM [1] as the Basic Medium

In this third experiment, explants were isolated in 10 different treatments described in Table 3. Nine treatments were used with different amounts ($1\times$, $0.5\times$ or $1.5\times$) of NH_4NO_3 and $Ca(NO_3)_2 \cdot 4H_2O$, using the original WPM [18] as the basic medium, and one treatment used the modified WPM [15] (Table 1). Media were supplemented as described above, with the addition of cytokinin ZEA (5 µM).

Table 3. Experiment 3 with 10 treatments. Treatments 1 to 9 with the original Woody Plant Medium (WPM) [18] with different amounts of NH_4NO_3 (x) and $Ca(NO_3)_2 \cdot 4H_2O$ (x) and treatment 10 with the modified WPM [15].

Treatments	1	2	3	4	5	6	7	8	9	10
NH_4NO_3 (x)	1×	1×	1×	1.5×	1.5×	1.5×	0.5×	0.5×	0.5×	-
$Ca(NO_3)_2 \cdot 4H_2O$ (x)	1×	0.5×	1.5×	1×	0.5×	1.5×	1×	0.5×	1.5×	-
Components	**Final Concentration in the Culture Medium (mg L^{-1})**									
NH_4NO_3	400.0	400.0	400.0	600.0	600.0	600.0	200.0	200.0	200.0	-
$Ca(NO_3)_2 \cdot 4H_2O$	556.0	278.0	834.0	556.0	278.0	834.0	556.0	278.0	834.0	278.0

2.3. Growing Conditions

After isolation, cultures were transferred to a culture room and grown at 25 °C $\pm$ 2 °C in the dark for eight initial days and then transferred to a 16-h photoperiod with a light intensity of 40 µmol m^{-2} s^{-1} provided by cool-day fluorescent lamps.

2.4. Experimental Design, Data Collection, and Statistical Analysis

The experiments were conducted in a completely randomized design. In experiment 1, a two-factor experiment (4×8) design was used, with four different cytokinins (ZEA, 2iP, BAP, and KIN) in eight different concentrations (0, 2.5, 5, 10, 20, 30, 40, and 50 µM). There were 32 treatments in total. Each treatment had four replicates of 10 tubes each (one plant per tube), e.g., 40 plants per treatment, resulting in a total of 1280 plants.

In experiment 2, a completely randomized design was used, with nine treatments, according to Table 2. Each treatment had three replicates of seven tubes each (one plant per tube), e.g., 21 plants per treatment, resulting in 189 plants.

In experiment 3, a completely randomized design was used, with 10 treatments (Table 3). Each treatment had four replicates of 10 tubes each (one plant per tube), e.g., 40 plants per treatment, resulting in a total of 400 plants.

Plants were evaluated based on many aspects two months (Experiment 1) or three months (Experiments 2 and 3) after initial culture. Contaminated cultures were discarded and not included in the data analysis. Contamination rates ranged from 0 to 7.5% in experiment 1. The final number of explants evaluated is presented in Table S1. In experiment 2, contamination rates ranged from 0 to 14%; and were 0 to 35% in experiment 3. Survival rate (%) and new shoot growth (%) were recorded. The number of new shoots formed per explant was counted (n°), the length of the longest shoot (millimeters from base to shoot tip) was measured, and the number of leaves of the longest shoot was counted (n°). All the plants were evaluated and had the mean estimated from the plants in each replication, and subsequently, the mean of the three or four replications in each treatment.

In experiment 1, ANOVA, Tukey, and regression analyses did not include values for the zero concentration treatments, since it was clear that a zero concentration did not have any influence on the explant development and it is not a concentration that labs would apply in practice. In the zero concentrations, there was no shoot formation in any of the explants evaluated. Since there was no shoot formation, there was no valid evaluation of number of shoots formed, length of shoot, or number of leaves per shoot. Hence, 28 treatments were statistically analyzed using a two-factor experiment (4 × 7), with four different cytokinins (ZEA, 2iP, BAP, and KIN) at seven different concentrations (2.5, 5, 10, 20, 30, 40, and 50 μM). The results were first transformed to the square root scale and then two-way ANOVA was performed (Table S2) to detect any interaction between the two factors and to check for any statistically significant difference between treatments at levels 1 and 5%. In the case of interaction between factors, in the variable analyzed, two tests were performed. First, Tukey's test ($P < 0.05$) was performed for each of the cytokinins with each of the concentrations. For factor 2 (different concentrations), regression analysis was performed for each cytokinin with the original data. The best-fitting regression model was obtained and the R^2 value was recorded. In experiments 2 and 3, original data were used, and one-way ANOVA was performed to check for any statistically significant difference between treatments ($P < 0.01$). Then, Scott-Knott's test ($P < 0.05$) was performed. For these analyses, the software Assistat® was used.

3. Results

3.1. Experiment 1: Cytokinins

In all the dependent variables analyzed (survival, shoot formation, number of shoots, length of shoot, and number of leaves), there was a significant interaction (at least $P < 0.01$) between the two factors (growth regulator and concentrations) tested, indicating that their effects are not independent. In addition, there was a significant difference between the different kinds of cytokinin tested for all the variables evaluated. F values were significant (at least ($P < 0.01$)) concerning factor 1 (different cytokinins) and concerning the interaction of factor 1 (different cytokinins) with factor 2 (different concentrations). Tukey's test results are shown in Table 4. The overall development of the explants in different cytokinin concentrations can be observed in Figure 1. The use of kinetin in the culture medium did not lead to any response in new shoots formed.

Table 4. Experiment 1, treatments with four cytokinins at different concentrations, showing mean values of survival (%), shoot formation (%), number of shoots (n°), length of shoot (mm), and number of leaves (n) in initial in vitro shoot culture of 'Delite' rabbiteye blueberry.

	Survival (%)						
Cytokinin	2.5 μM	5 μM	10 μM	20 μM	30 μM	40 μM	50 μM
ZEA	89.7 ± 14.2a [z]	96.4 ± 7.1a	92.2 ± 5.2a	94.7 ± 6.1a	100.0 ± 0.8a	100.0 ± 0.8a	100.0 ± 0.8a
2iP	36.9 ± 14.4b	60.0 ± 20.0b	78.1 ± 11.4a	78.6 ± 16.0a	94.7 ± 6.1a	100.0 ± 0.8a	100.0 ± 0.8a
BAP	24.2 ± 11.6b	52.5 ± 6.8b	59.3 ± 8.3a	82.2 ± 16.9a	71.9 ± 14.7a	68.9 ± 10.1a	73.6 ± 21.6a
KIN	2.8 ± 5.6c	0.0 ± 0.1c	5.0 ± 5.8b	5.3 ± 6.1b	0.0 ± 0.1b	8.3 ± 5.6b	7.8 ± 6.1b
Mean	38.4	52.2	58.7	65.2	66.7	69.3	70.4

	Shoot Formation (%)						
Cytokinin	2.5 μM	5 μM	10 μM	20 μM	30 μM	40 μM	50 μM
ZEA	81.3 ± 9.2a	88.2 ± 1.8a	90.0 ± 0.1a	94.7 ± 6.1a	100.0 ± 0.0 a	100.0 ± 0.0a	100.0 ± 0.0a
2iP	0.0 ± 0.0b	0.0 ± 0.0b	30.6 ± 22.4b	42.2 ± 8.6b	53.3 ± 14.4b	70.0 ± 12.4b	95.0 ± 5.8a
BAP	0.0 ± 0.0b	0.0 ± 0.0b	0.0 ± 0.0c	0.0 ± 0.0c	5.0 ± 5.8c	7.8 ± 5.8c	7.5 ± 5.0b
KIN	0.0 ± 0.0b	0.0 ± 0.0b	0.0 ± 0.0c	0.0 ± 0.0c	0.0 ± 0.0d	0.0 ± 0.0d	0.0 ± 0.0c
Mean	20.3	22.1	30.2	34.2	39.6	44.4	50.6

	Number of Shoots per Explant (n°)						
Cytokinin	2.5 μM	5 μM	10 μM	20 μM	30 μM	40 μM	50 μM
ZEA	1.3 ± 0.1a	1.4 ± 0.1a	1.3 ± 0.1a	1.5 ± 0.1a	1.3 ± 0.1a	1.5 ± 0.1a	1.6 ± 0.1a
2iP	0.0 ± 0.0b	0.0 ± 0.0b	1.0 ± 0.0a	1.1 ± 0.1a	1.0 ± 0.0a	1.5 ± 0.1a	1.2 ± 0.1ab
BAP	0.0 ± 0.0b	0.0 ± 0.0b	0.0 ± 0.0b	0.0 ± 0.0b	0.5 ± 0.6b	0.8 ± 0.5b	0.8 ± 0.5b
KIN	0.0 ± 0.0b	0.0 ± 0.0b	0.0 ± 0.0b	0.0 ± 0.0b	0.0 ± 0.0c	0.0 ± 0.0c	0.0 ± 0.0c
Mean	0.3	0.3	0.6	0.7	0.7	0.9	0.9

	Shoot Length (mm)						
Cytokinin	2.5 μM	5 μM	10 μM	20 μM	30 μM	40 μM	50 μM
ZEA	13.8 ± 3.4a	8.4 ± 1.1a	5.6 ± 1.8a	6.7 ± 3.4a	4.1 ± 0.3a	3.9 ± 0.6ab	5.0 ± 0.8a
2iP	0.0 ± 0.0b	0.0 ± 0.0b	3.0 ± 0.8a	3.3 ± 0.3b	3.9 ± 0.4a	4.7 ± 0.6a	4.2 ± 0.4a
BAP	0.0 ± 0.0b	0.0 ± 0.0b	0.0 ± 0.0b	0.0 ± 0.0c	1.3 ± 1.5b	2.5 ± 2.1b	1.8 ± 1.3b
KIN	0.0 ± 0.0b	0.0 ± 0.0b	0.0 ± 0.0b	0.0 ± 0.0c	0.0 ± 0.0c	0.0 ± 0.0c	0.0 ± 0.0c
Mean	3.4	2.1	2.1	2.5	2.3	2.8	2.7

	Number of Leaves (n°)						
Cytokinin	2.5 μM	5 μM	10 μM	20 μM	30 μM	40 μM	50 μM
ZEA	10.0 ± 1.3a	8.8 ± 1.3a	6.3 ± 2.3a	8.0 ± 3.2a	6.0 ± 0.9a	6.1 ± 1.5a	7.9 ± 1.1a
2iP	0.0 ± 0.0b	0.0 ± 0.0b	2.4 ± 1.1b	3.0 ± 1.3b	3.2 ± 0.3b	5.7 ± 1.0a	5.8 ± 0.7a
BAP	0.0 ± 0.0b	0.0 ± 0.0b	0.0 ± 0.0c	0.0 ± 0.0c	0.8 ± 0.0c	1.0 ± 0.0b	0.5 ± 0.0b
KIN	0.0 ± 0.0b	0.0 ± 0.0b	0.0 ± 0.0c	0.0 ± 0.0c	0.0 ± 0.0c	0.0 ± 0.0c	0.0 ± 0.0b
Mean	2.5	2.2	2.2	2.7	2.5	3.2	3.6

[z] Data are the means of four replicates ± standard deviation (SD). Means followed by the same lowercase letter within the column are not significantly different according to Tukey's test ($P < 0.05$).

Figure 1. Initial in vitro shoot culture of 'Delite' rabbiteye blueberry in eight different concentrations (0, 2.5, 10, 20, 30, 40, and 50 µM) with four different cytokinins: (**a**) ZEA, (**b**) 2iP, (**c**) BAP, and (**d**) KIN. Bars represent 2 cm. Abbreviations: BAP, 6-benzylaminopurine; KIN, kinetin; ZEA, zeatin; 2iP, 6-(γ-γ-dimethylallylamino)-purine.

3.1.1. The Effects of Cytokinins on Survival

ZEA was superior to the other cytokinins at the concentrations of 2.5 and 5 µM. In all concentrations, KIN had the worst performance for survival rate. Finally, in the concentration range of 10–50 µM, ZEA, 2iP, and BAP all had the same effect on survival. The regression analyses can be observed in Figure 2.

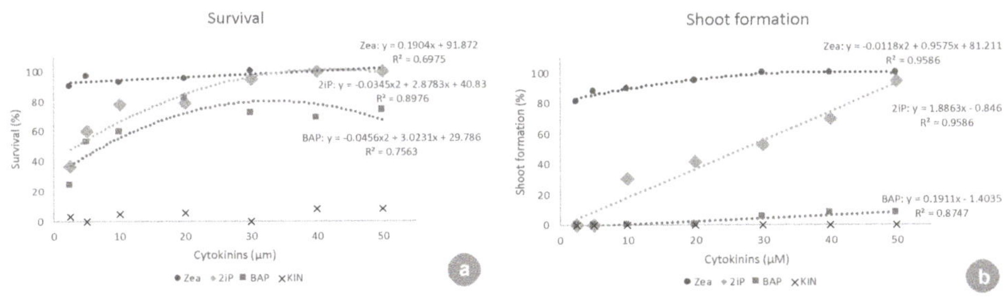

Figure 2. *Cont.*

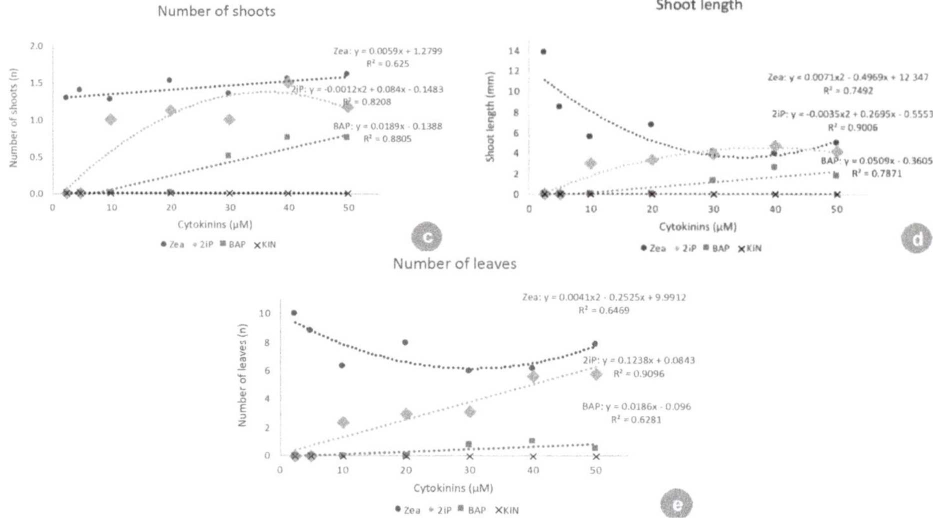

Figure 2. Regression analysis related to different cytokinin (ZEA, 2iP, BAP, and KIN) concentrations (2.5, 5, 10, 20, 30, 40, and 50 μM) on in vitro establishment of 'Delite' rabbiteye blueberry. (**a**) Survival (%); (**b**) shoot formation (%); (**c**) number of shoots (n°); (**d**) shoot length (mm); (**e**) number of leaves (n°). Abbreviations: BAP, 6-benzylaminopurine; KIN, kinetin; ZEA, zeatin; 2iP, 6-(γ-γ-dimethylallylamino)-purine.

3.1.2. The Effects of Cytokinins on Shoot Formation

Shoot formation from the initial explant was highly influenced by different cytokinins. According to the quadratic polynomial regression analysis across ZEA concentrations (Figure 2), a maximum shoot formation of 100% would be acquired at a concentration of 40.6 μM.

The evaluation of different means can be observed in Table 4, where in almost all of the concentrations tested (except 50 μM), ZEA was superior to all the other treatments, varying from 81.3 to 100% shoot formation. At concentrations of 2.5 and 5 μM, 2iP, BAP, and KIN did not show any response. 2iP showed responses from 10 to 50 μM only, presenting a rate varying from 30.6 to 95.0% in those concentrations. In concentrations of 10, 20, 30, and 40 μM, 2iP was the second cytokinin to form shoots. At a concentration of 50 μM, 2iP was equivalent to ZEA, and both were superior to BAP and KIN in this concentration. BAP did not show any response in the explants growing at the lowest concentrations of 2.5, 5, 10, and 20 μM. The first response for BAP appeared only at the concentrations of 30, 40, and 50 μM, showing a rate of shoot formation of only 5.0 to 7.8% of explants showing new shoot formation. BAP had lower shoot formation than ZEA and 2iP at all the concentrations tested.

3.1.3. The Efects of Cytokinins on the Number of Shoots Per Explant

Regarding the number of new shoots formed, we can observe that ZEA showed a linear relationship (Figure 2) and calculate that a concentration of 22.0 μM would be required to reach 1.4 shoots per explant. With 2iP, the maximum point in the curve reached 1.4 shoots per explant, which would be acquired at a concentration of 37.37 μM 2iP.

At the concentrations of 2.5 and 5 μM (Table 4), ZEA was superior to all the other cytokinins, showing 1.3 and 1.4 shoots per explant, respectively. At concentrations of 10, 20, 30, 40, and 50 μM, ZEA and 2iP had the same performance and were superior to BAP and kinetin. BAP only showed some shoot formation at concentrations of 30, 40, and 50 μM, exhibiting an average of only 0.5 to 0.8 new shoots per explant.

3.1.4. The Effects of Cytokinins on Shoot Length

Type of cytokinin had a significant influence on shoot length. Regression analysis (Figure 2) shows that ZEA followed a quadratic polynomial trend, with the concavity upward, showing an initial higher shoot length in the lowest concentrations (11.1 mm calculated at 2.5 µM), decreasing to the lowest point (3.6 mm) at 34.8 µM ZEA, and then starting to increase again. 2iP had a quadratic polynomial trend with the concavity downward. The maximum point in this curve was 4.56 mm of shoot length at 38.2 µM 2iP.

At the lowest concentrations of growth regulators, 2.5 and 5 µM ZEA was superior to all the other treatments, showing shoots with 13.8 and 8.4 mm, respectively (Table 4). In these two concentrations, 2iP, BAP, and KIN did not show any new shoots. At the other concentrations tested, 10, 20, 30, 40, and 50 µM, 2iP showed new shoots. At concentrations of 10, 30, 40, and 50 µM, 2iP treatments presented shoot lengths that did not differ from those of ZEA; ZEA and 2iP displayed an equal performance. BAP showed smaller shoots compared to 2iP and ZEA at all the concentrations except 40 µM. BAP only showed new shoots at the concentrations of 30 µM (1.3 mm), 40 µM (2.5 mm), and 50 µM (1.8 mm). Kinetin was inferior to all the others, in all the concentrations tested.

3.1.5. The Effects of Cytokinins on Number of Leaves

The number of leaves was significantly influenced by different cytokinins. The ZEA regression curve was a quadratic polynomial with concavity upward (Figure 2), similar to the curve observed for the influence of ZEA concentrations on shoot length. The minimum value in this curve was 6.2 leaves, reached at the concentration of 30.7 µM ZEA. 2iP behaved in a linear way, showing a maximum of 5.8 leaves at a concentration of 50 µM. BAP was also represented by a linear relationship, reaching a maximum of 0.8 leaves at the highest concentration of 50 µM.

At concentrations of 2.5, 5, 10, 20, and 30 µM, ZEA was superior to all the other cytokinins, showing 10.0, 8.8, 6.3, 8.0, and 6.0 leaves per shoot (Table 4). 2iP was inferior to ZEA at all concentrations, except at the highest concentrations of 40 and 50 µM, where both cytokinins were equivalent. BAP was always inferior to ZEA. At concentrations of 10, 20, 30, 40, and 50 µM, BAP was also inferior to 2iP. BAP only showed some leaves at concentrations of 30 µM (0.8 leaves), 40 µM (1.0 leaf), and 50 µM (0.5 leaves). At all of the concentrations, KIN did not show any response.

3.2. Experiment 2: Combinations of $(NH_4)_2SO_4$, KNO_3, and $Ca(NO_3)_2 \cdot 4H_2O$ Using the Modified WPM [15] as Basic Medium

There were no statistically significant differences among the nine treatments tested for any of the variables analyzed. Survival and shoot formation rates ranged from 43.7 to 76.2%, the number of shoots formed ranged from 1.1 to 1.4, shoot lengths ranged from 7.5 to 25.0 mm, and the number of leaves ranged from 9.4 to 19.7 (Table 5).

Table 5. Experiment 2 with treatments 1 to 9 on the modified Woody Plant Medium (modified WPM) [15] showing the number of explants evaluated (n°), survival rate (%), shoot formation (%), number of shoots (n°), shoot length (mm), and number of leaves (n°) in 'Delite' rabbiteye blueberry in vitro establishment. Abbreviations: CV, coefficient of variation; n°, number.

Treatment	$(NH_4)_2SO_4$ (x)	KNO_3 and $Ca(NO_3)_2 \cdot 4H_2O$ (x)	n°	Survival (%)	Shoot Formation (%)	Number of Shoots (n°)	Shoot Length (mm)	Number sf Leaves (n°)
1-modified WPM	1×	1×	19	48.4 ± 19.1a [z]	48.4 ± 19.1a	1.4 ± 0.1a	13.9 ± 6.7a	11.1 ± 3.4a
2	1×	0.5×	19	62.7 ± 11.3a	62.7 ± 11.3a	1.3 ± 0.2a	24.9 ± 7.4a	19.7 ± 3.1a
3	1×	1.5×	20	43.7 ± 23.4a	43.7 ± 23.4a	1.3 ± 0.3a	17.7 ± 16.9a	13.7 ± 7.6a
4	0.5×	1×	18	73.0 ± 35.1a	73.0 ± 35.1a	1.1 ± 0.2a	9.1 ± 5.5a	10.9 ± 5.6a
5	0.5×	0.5×	19	69.5 ± 11.5a	69.5 ± 11.5a	1.1 ± 0.1a	10.7 ± 3.2a	11.8 ± 1.5a
6	0.5×	1.5×	20	45.2 ± 4.1a	45.2 ± 4.1a	1.2 ± 0.2a	25.0 ± 14.3a	18.0 ± 6.0a
7	1.5×	1×	21	76.2 ± 8.2a	76.2 ± 8.2a	1.2 ± 0.0a	7.5 ± 2.7a	9.4 ± 3.3a
8	1.5×	0.5×	20	56.3 ± 23.4a	56.3 ± 23.4a	1.1 ± 0.2a	17.2 ± 9.7a	12.7 ± 1.9a
9	1.5×	1.5×	20	54.0 ± 19.2a	54.0 ± 19.2a	1.4 ± 0.4a	21.2 ± 7.7a	16.8 ± 5.0a
Mean	-	-	-	58.8	58.8	1.2	16.4	13.8
CV%	-	-	-	33.1	33.1	16.5	57.4	33.1

[z] Data are the means of three replicates ± standard deviation (SD). Means followed by the same lowercase letter within a column are not significantly different according to Scott-Knott's test ($P < 0.05$).

3.3. Experiment 3: Combinations of NH_4NO_3 and $Ca(NO_3)_2 \cdot 4H_2O$ Using the Original WPM [18] as the Basic Medium

In this experiment, it was possible to verify that treatments 7 ($0.5 \times NH_4NO_3$ and $1 \times Ca(NO_3)_2$), 8 ($0.5 \times NH_4NO_3$ and $0.5 \times Ca(NO_3)_2$), and 10 (modified WPM) showed the lowest rates of survival and shoot formation and shortest shoot length (Table 6).

Table 6. Experiment 3 with treatments 1 to 9 on Woody Plant Medium (WPM) [18] with different ranges of NH_4NO_3 (x) and $Ca(NO_3)_2 \cdot 4H_2O$ (x) and treatment 10 on modified WPM [15], showing number of explants evaluated (n°), survival rate (%), shoot formation (%), number of shoots (n°), shoot length (mm), and number of leaves (n°) in 'Delite' rabbiteye blueberry in vitro establishment. Abbreviations: CV, coefficient of variation; n°, number.

Treatment	Solution NH_4NO_3	Solution $Ca(NO_3)_2$	n°	Survival (%)	Shoot Formation (%)	Number of Shoots (n°)	Shoot Length (mm)	Number of Leaves (n°)
1-original WPM	1×	1×	38	79.4 ± 16.4a [z]	79.4 ± 16.4a	1.2 ± 0.3a	33.3 ± 7.6a	15.3 ± 2.1a
2	1×	0.5×	40	95.0 ± 5.8a	95.0 ± 5.8a	1.1 ± 0.1a	23.5 ± 3.9b	12.2 ± 1.5a
3	1×	1.5×	38	97.5 ± 5.0a	97.5 ± 5.0a	1.2 ± 0.1a	21.0 ± 1.9b	11.4 ± 1.7a
4	1.5×	1×	27	86.8 ± 10.5a	83.7 ± 15.7a	1.1 ± 0.1a	25.4 ± 8.9b	12.1 ± 2.9a
5	1.5×	0.5×	40	92.5 ± 5.0a	90.0 ± 8.2a	1.1 ± 0.2a	32.3 ± 7.3a	14.4 ± 1.1a
6	1.5×	1.5×	40	90.0 ± 8.2a	90.0 ± 8.2a	1.1 ± 0.1a	30.7 ± 6.6a	14.1 ± 1.2a
7	0.5×	1×	39	64.4 ± 16.7b	64.4 ± 16.7b	1.1 ± 0.1a	5.0 ± 0.7d	6.6 ± 1.2c
8	0.5×	0.5×	35	55.1 ± 5.6b	55.1 ± 5.6b	1.1 ± 0.1a	18.2 ± 10.5c	10.4 ± 3.6b
9	0.5×	1.5×	26	85.4 ± 17.2a	85.4 ± 17.2a	1.0 ± 0.0a	24.2 ± 3.8b	14.9 ± 1.6a
10-modified WPM			32	55.7 ± 21.4b	55.7 ± 21.4b	1.2 ± 0.2a	15.7 ± 1.3c	12.9 ± 2.2a
Mean	-	-	-	80.2	79.6	1.1	22.9	12.4
CV%	-	-	-	16.4	7.4	15.0	26.9	16.6

[z] Data are the means of four replicates ± standard deviation (SD). Means followed by the same lowercase letter within the column are not significantly different according to Scott-Knott's Test ($P < 0.05$).

The number of shoots was similar in all the treatments tested. In addition, concerning the number of leaves, the lowest number was obtained with treatments 7 and 8. Observing survival, shoot formation, and shoot length, treatment 1 (original WPM) was superior to treatment 10 (modified WPM) (Figure 3).

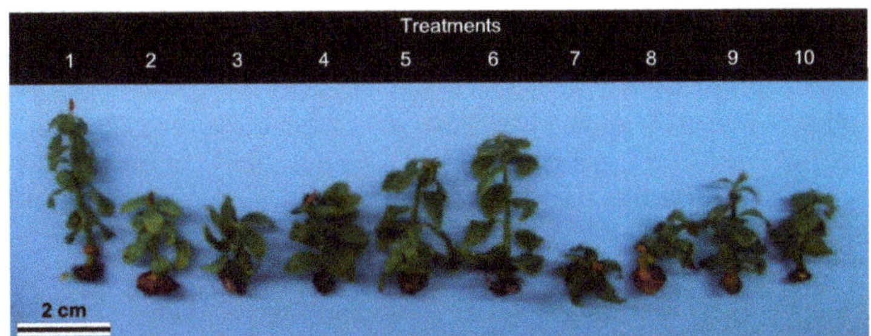

Figure 3. Initial in vitro shoot culture of 'Delite' rabbiteye blueberry with 10 treatments. Treatments 1 to 9 with the original Woody Plant Medium (WPM) [18] with different ranges of NH_4NO_3 (x) and $Ca(NO_3)_2 \cdot 4H_2O$ (x), compared to treatment 10 (modified Woody Plant Medium [15]).

4. Discussion

In vitro establishment is an important step in tissue culture. It is a critical point where explants come from a different environment and have to adapt to in vitro conditions. One of the key steps in this process is the use of adequate growth regulators and a balance of mineral salts in a suitable concentration. Our results showed a screening comparison of four different cytokinins in eight different concentrations and varying balances of nitrogen salts in 'Delite' rabbiteye blueberry, presenting an efficient technique for in vitro plant propagation of this species.

The species and cultivars of the *Vaccinium* genus show natural variation in in vitro responses. There is high genetic variation in growth regulator responses/needs. Our results, based on linear and quadratic polynomial regression analyses, displayed the effects of cytokinin concentrations and their great impact on the survival of explants, new shoot formation, number of new shoots formed, length of shoots formed, and number of leaves in the shoots.

ZEA and 2iP resulted in better responses to in vitro establishment. At the lowest concentrations tested, 2.5 and 5 µM, ZEA was superior to all of the other cytokinins tested, in all the variables analyzed, presenting values of: 89.7 and 96.4% explant survival, 81.3 and 88.2% of explants forming new shoots, 1.3 and 1.4 new shoots formed, 13.8 and 8.4 mm of shoot length, and 10.0 and 8.8 leaves per shoot, respectively. Similar results were observed with highbush blueberry 'Polaris' and half-high blueberry 'St. Cloud', where ZEA was used at a concentration of 9.1 µM in the shoot establishment in vitro. ZEA was also efficient in inducing shoot proliferation in a liquid medium at 4.6 µM [10], instead of at higher concentrations. For *V. corymbosum* 'Oskar', *V. angustifolium* 'Emil' and 'Putte', and *V. corymbosum* × *V. angustifolium* 'Northblue' establishment, 2 mg L^{-1} (9.12 µM) ZEA was used [20]. In highbush blueberry 'Duke' propagation, ZEA at 2 mg L^{-1} (9.12 µM) was superior to 2iP or TDZ [21].

For in vitro shoot proliferation in cranberry (*V. macrocarpon* Ait.) cultivars, ZEA at very low concentrations (2–4 µM) showed good results [22]. In *V. ashei* at the multiplication stage, ZEA increased shoot formation compared to 2iP. However, 2iP showed longer shoots with a higher number of nodes [23]. For initial culture of highbush blueberry, 1 mg L^{-1} (2.85 µM) zeatin riboside was used [24]. In lowbush blueberry, the authors tested 0, 2.3, 4.6, or 9.10 µM ZEA on the elongation of shoots, and concentrations of 2.3 and 4.6 µM gave the best response [25].

Another important aspect is the growth habit of the 'Delite' rabbiteye blueberry cultivar in this study. In particular, in the presence of ZEA and 2iP, it showed a low number of new shoots per explant, but longer shoots, which means that a new subculture could be performed using the nodal segments of the long shoot instead of using new axillary or adventitious shoots formed.

At the lowest concentrations (2.5 and 5 µM), there were no responses to 2iP. Treatments with 2iP started to form shoots only at the concentrations of 10, 20, 30, 40, and 50 µM. Concerning the percentage of explants forming new shoots, 2iP was inferior to ZEA in all of the concentrations, except 50 µM, where both had the same shoot formation rate. This showed that ZEA triggered a response in the explants, even at inferior concentrations (2.5 and 5 μM), and that 2iP was able to lead to some shoot formation only at higher concentrations (10 μM and above). Concerning shoot length, at the concentrations where 2iP started showing new shoots (10–50 μM), the shoots formed were equivalent in length to the shoots formed with ZEA. At concentrations of 10, 20, 30, and 50 μM, both were superior to BAP and KIN. However, when analyzing the number of leaves, ZEA was superior to 2iP at almost all concentrations, except 40 and 50 μM, again demonstrating the need for higher concentrations of 2iP to produce a higher number of leaves. In 'Brightwell' blueberry, the authors found that different concentrations of 2iP (5, 10, 15, or 20 mg·L^{-1}) and TDZ were inferior to 2 mg L^{-1} (9.12 μM) ZEA in shoot proliferation [26]. ZEA at 4 mg·L^{-1} (18.24 μM) was more successful than 2iP at 10 or 15 mg·L^{-1} (49.2 or 73.8 μM) in establishing *V. corymbosum* blueberry cultivars [12].

BAP did not show any response at the lowest concentrations of 2.5, 5, 10, and 20 μM. BAP only started showing a low response to 30, 40, and 50 μM (5.0–7.8% shoot formation). BAP was always highly inferior to ZEA at all concentrations tested, in all of the variables analyzed, except for the shoot length at 40 μM. Additionally, BAP was inferior to 2iP from 10–50 μM concerning shoot formation, number of shoots, shoot length, and number of leaves. In the same way, in 'Bluejay' and 'Pink Lemonade' blueberry, the authors found that BAP induced fewer axillary shoots than ZEA, as well as smaller shoots [27].

Kinetin showed no response concerning shoot formation and had almost no surviving explants (maximum of 8.3% survival), clearly showing that it was not suitable for 'Delite' rabbiteye blueberry initiation culture.

In this study, different balances of nitrogen salts were tested. Using the modified WPM medium, no differences were observed among all combinations of nitrogen salts: $(NH_4)_2SO_4$, KNO_3, and $Ca(NO_3)_2 \cdot 4H_2O$. 'Delite' blueberry showed lower survival (55.7%), shoot formation (55.7%), and shoot length (15.7 mm) in the modified WPM compared with the original WPM (79.4%, 79.4%, and 33.3 mm, respectively).

Using the original WPM, it was observed that treatments containing higher amounts of NH_4NO_3 (1$\times$ or 1.5$\times$, instead of 0.5$\times$), as well as the treatment with a higher amount of $Ca(NO_3)_2$ (1.5$\times$), even with a lower amount of NH_4NO_3 (0.5$\times$), showed the same performance as in WPM without modification. Similarly, in red raspberries, it was found that combinations of intermediate to high NO_3^- and intermediate to high NH_4^+ developed the most growth in most cultivars [28].

However, changing the ranges of $Ca(NO_3)_2$, in addition to increasing or decreasing the total amount of nitrogen and its nitrate form, would also change the Ca^{+2} ion. Therefore, the result seen in the treatment $Ca(NO_3)_2$ (1.5$\times$) could be related to either nitrogen or calcium in higher amounts, or even both.

This study in a rabbiteye blueberry cultivar represents a basic framework that can be used to understand initial in vitro establishment. It can be useful to describe this process in other *Vaccinium* cultivars regarding the adjustments necessary to adapt the process to different genotypes.

5. Conclusions

This research showed the effects of different cytokinins at different concentrations and different nitrogen salt ranges on 'Delite' rabbiteye blueberry during in vitro establishment, and it provides basic knowledge for further experiments in rabbiteye blueberry tissue culture.

In conclusion, focusing on an efficient strategy for in vitro establishment of 'Delite' rabbiteye blueberry, we recommend the lowest concentration tested, 2.5 μM ZEA, which promoted a high survival rate (89.7%), as well as a good response on explants forming new shoots (81.3%). This concentration yielded a number of 1.3 new shoots, a high shoot length (13.8 mm), and 10.0 leaves per shoot. Concerning salt composition, we recommend the original WPM. An increase or decrease in the NH_4NO_3 and $Ca(NO_3)_2$ concentration did not promote better growth than the original medium.

This work is of interest for evaluating different cytokinin and salt compositions in the culture medium for in vitro establishment of rabbiteye blueberry, and it can contribute to developing a deeper knowledge of large-scale propagation, germplasm conservation, and development of other biotechnology techniques in other research fields, such as morphology, plant breeding, and physiology.

Future studies could be developed beyond the research presented here, focusing on fine-tuning the salts composition and concentrations of the growth regulator needed for an efficient response, as well as combining the two most successful cytokinins tested, ZEA and 2iP.

Supplementary Materials: The following are available online at http://www.mdpi.com/2311-7524/5/1/24/s1, Table S1: Results of the two-way ANOVA of experiment 1 studying the influence of cytokinin type and concentration on survival (%), shoot formation (%), number of shoots (n°), length of shoot (mm), and number of leaves (n°) on initial in vitro shoot culture of 'Delite' rabbiteye blueberry, Table S2: Number of explants evaluated after contamination in experiment 1 in each of the treatments (cytokinin type by concentration) on initial in vitro shoot culture of 'Delite' rabbiteye blueberry.

Author Contributions: C.S.S. contributed the design of the study, ran the laboratory work, processed the data, and drafted the paper. L.A.B. supervised the research and provided critical reading for the final version of the manuscript.

Funding: This research was funded in part by Coordenação de Aperfeiçoamento de Pessoal de Nível Superior—Brasil (CAPES)—Finance Code 001 under a PhD scholarship grant.

Acknowledgments: We are very grateful to Marino Schiehl and Kassius Augusto da Rosa for their support and collaboration in the laboratory work.

Conflicts of Interest: The authors declare no conflict of interest.

Abbreviations

BAP	6-Benzylaminopurine
CV	coefficient of variation
DF	degrees of freedom
KIN	kinetin: 6-furfurylaminopurine
MS	mean squares
SS	sum of squares
WPM	Woody Plant Medium
ZEA	zeatin: 6-(4-Hydroxy-3-methylbut-2-enylamino)purine
2iP	6-(γ-γ-dimethylallylamino)-purine

References

1. Retamales, J.B.; Hancock, J.F. *Blueberries*; CABI: Oxfordshire, UK, 2012.
2. Michalska, A.; Łysiak, G. Bioactive compounds of blueberries: Post-harvest factors influencing the nutritional value of products. *Int. J. Mol. Sci.* **2015**, *16*, 18642–18663. [CrossRef] [PubMed]
3. Poulose, S.M.; Miller, M.G.; Scott, T.; Shukitt-Hale, B. Nutritional factors affecting adult neurogenesis and cognitive function. *Adv. Nutr. Int. Rev. J.* **2017**, *8*, 804–811. [CrossRef] [PubMed]
4. Cheng, A.; Yan, H.; Han, C.; Wang, W.; Tian, Y.; Chen, X. Polyphenols from blueberries modulate inflammation cytokines in LPS-induced RAW264.7 macrophages. *Int. J. Biol. Macromol.* **2014**, *69*, 382–387. [CrossRef] [PubMed]
5. Balducci, F.; Capocasa, F.; Mazzoni, L.; Mezzetti, B.; Scalzo, J. Study on adaptability of blueberry cultivars in center-south Europe. *Acta Hortic.* **2016**, 53–58. [CrossRef]
6. Debnath, S.C. A scale-up system for lowbush blueberry micropropagation using a bioreactor. *HortScience* **2009**, *44*, 1962–1966. [CrossRef]
7. Routray, W.; Orsat, V. Blueberries and their anthocyanins: Factors affecting biosynthesis and properties. *Compr. Rev. Food Sci. Food Saf.* **2011**, *10*, 303–320. [CrossRef]
8. Marino, S.R.; Williamson, J.G.; Olmstead, J.W.; Harmon, P.F. Vegetative growth of three southern highbush blueberry cultivars obtained from micropropagation and softwood cuttings in two Florida locations. *HortScience* **2014**, *49*, 556–561. [CrossRef]
9. Meiners, J.; Schwab, M.; Szankowski, I. Efficient in vitro regeneration systems for *Vaccinium* species. *Plant Cell Tissue Organ Cult.* **2007**, *89*, 169–176. [CrossRef]

10. Debnath, S.C. Temporary immersion and stationary bioreactors for mass propagation of true-to-type highbush, half-high, and hybrid blueberries (*Vaccinium* spp.). *J. Hortic. Sci. Biotechnol.* **2017**, *92*, 72–80. [CrossRef]

11. Goyali, J.C.; Igamberdiev, A.U.; Debnath, S.C. Propagation methods affect fruit morphology and antioxidant properties but maintain clonal fidelity in lowbush blueberry. *HortScience* **2015**, *50*, 888–896. [CrossRef]

12. Reed, B.M.; Abdelnour-Esquivel, A. The use of zeatin to initiate in vitro cultures of *Vaccinium* species and cultivars. *HortScience* **1991**, *26*, 1320–1322. [CrossRef]

13. Jaakola, L.; Tolvanen, A.; Laine, K.; Hohtola, A. Effect of N6-isopentenyladenine concentration on growth initiation in vitro and rooting of bilberry and lingonberry microshoots. *Plant Cell Tissue Organ Cult.* **2001**, *66*, 73–77. [CrossRef]

14. Tetsumura, T.; Matsumoto, Y.; Sato, M.; Honsho, C.; Yamashita, K.; Komatsu, H.; Sugimoto, Y.; Kunitake, H. Evaluation of basal media for micropropagation of four highbush blueberry cultivars. *Sci. Hortic.* **2008**, *119*, 72–74. [CrossRef]

15. Wang, Y.; Wang, S.; Xu, J.; Yan, Z.; Bao, H. Culture Medium for Proliferating Blueberry Tissue, Comprises Improved Woody Plant Medium Culture Medium, Naphthaleneacetic Acid, Gibberellic Acid, Zeatin, Sucrose and Agar. Patent Number CN104082142-A; CN104082142-B, 8 October 2014.

16. Reed, B.M.; Wada, S.; DeNoma, J.; Niedz, R.P. Mineral nutrition influences physiological responses of pear in vitro. *In Vitro Cell. Dev. Biol. Plant* **2013**, *49*, 699–709. [CrossRef]

17. Ramage, C.M.; Williams, R.R. Inorganic nitrogen requirements during shoot organogenesis in tobacco leaf discs. *J. Exp. Bot.* **2002**, *53*, 1437–1443. [CrossRef] [PubMed]

18. Lloyd, G.; McCown, B. Commercially-feasible micropropagation of mountain laurel, *Kalmia latifolia*, by use of shoot-tip culture. *Int. Plant Propagators Soc. Comb. Proc.* **1980**, *30*, 421–427.

19. Murashige, T.; Skoog, F. A revised medium for rapid growth and bio assays with tobacco tissue cultures. *Physiol. Plant.* **1962**, *15*, 473–497. [CrossRef]

20. Welander, M.; Sayegh, A.; Hagwall, F.; Kuznetsova, T.; Holefors, A. Technical improvement of a new bioreactor for large scale micropropagation of several *Vaccinium* cultivars. *Acta Hortic.* **2017**, 387–392. [CrossRef]

21. Cappelletti, R.; Sabbadini, S.; Mezzetti, B. The use of TDZ for the efficient in vitro regeneration and organogenesis of strawberry and blueberry cultivars. *Sci. Hortic.* **2016**, *207*, 117–124. [CrossRef]

22. Debnath, S.C. Zeatin-induced one-step in vitro cloning affects the vegetative growth of cranberry (*Vaccinium macrocarpon* Ait.) micropropagules over stem cuttings. *Plant Cell Tissue Organ Cult.* **2008**, *93*, 231–240. [CrossRef]

23. Schuch, M.W.; Damiani, C.R.; Silva, L.C.; Erig, A.C. Micropropagação como técnica de rejuvenescimento em mirtilo (*Vaccinium ashei* Reade) cultivar climax. *Ciênc. Agrotecnol.* **2008**, *32*, 814–820. [CrossRef]

24. Hung, C.D.; Hong, C.-H.; Kim, S.-K.; Lee, K.-H.; Park, J.-Y.; Nam, M.-W.; Choi, D.-H.; Lee, H.-I. LED light for in vitro and ex vitro efficient growth of economically important highbush blueberry (*Vaccinium corymbosum* L.). *Acta Physiol. Plant.* **2016**, *38*, 152. [CrossRef]

25. Debnath, S.C. A two-step procedure for adventitious shoot regeneration on excised leaves of lowbush blueberry. *In Vitro Cell. Dev. Biol. Plant* **2009**, *45*, 122–128. [CrossRef]

26. Jiang, Y.; Zhang, D.; He, S.; Wang, C. Influences of media and cytokinins on shoot proliferation of "Brightwell" and "Choice" blueberries in vitro. *Acta Hortic.* **2009**, *810*, 581–586. [CrossRef]

27. Fan, S.; Jian, D.; Wei, X.; Chen, J.; Beeson, R.C.; Zhou, Z.; Wang, X. Micropropagation of blueberry 'Bluejay' and 'Pink Lemonade' through in vitro shoot culture. *Sci. Hortic.* **2017**, *226*, 277–284. [CrossRef]

28. Poothong, S.; Reed, B.M. Optimizing shoot culture media for *Rubus* germplasm: The effects of NH_4^+, NO_3^-, and total nitrogen. *In Vitro Cell. Dev. Biol. Plant* **2016**, *52*, 265–275. [CrossRef]

 *horticulturae*

Communication

Pecan Propagation: Seed Mass as a Reliable Tool for Seed Selection

Tales Poletto [1,*], Valdir Marcos Stefenon [2], Igor Poletto [3] and Marlove Fátima Brião Muniz [1]

[1] Department of Plant Defense, Federal University of Santa Maria, Santa Maria 97105-900, Brazil; marlovemuniz@yahoo.com.br

[2] Nucleus of Molecular Ecology and Plant Micropropagation, Federal University of the Pampa, Rod. BR290, São Gabriel ZIP 97300-000, Brazil; valdirstefenon@unipampa.edu.br

[3] Laboratory of Biological Control and Plant Protection, Federal University of the Pampa, Rod. BR290, São Gabriel ZIP 97300-000, Brazil; igorpoletto@unipampa.edu.br

* Correspondence: tecnicotales@hotmail.com; Tel.: +55-55-99922-2060

Received: 26 August 2018; Accepted: 6 September 2018; Published: 11 September 2018

Abstract: Pecan is one of the most important horticultural nut crops in the world. It is a deciduous species native to the temperate zones of North America, introduced into the subtropical regions of Brazil during the 1870s. High quality seedlings are essential to establishing healthy and productive orchards, and selection of seeds is an important factor in this issue. In this study we evaluated the correlation between seed mass, emergence rate and morphometric traits of seedlings in the pecan cultivar *Importada*. A significant positive correlation ($r > 0.81$) between seed mass and plantlet height, stem diameter, emergence rate and number of leaves was observed. Our results suggest that seed mass can be used as a direct method for seed selection towards production of vigorous pecan seedlings. However, since an increase in seed mass is usually associated with a decrease in the number of seeds that a plant can produce per unit canopy, long-duration studies are recommended in order to evaluate the influence of seed selection on a plantation's production.

Keywords: *Carya illinoinensis*; orchards; seedlings production; emergence rate

1. Introduction

The pecan (*Carya illinoinensis* (Wangenh) K. Koch) is a deciduous tree species of the Junglandaceae family, native to the temperate zones of North America, and one of the most important horticultural nut crops in the world. Pecan is cultivated in a wide range of countries, including USA, Mexico, Israel, South Africa, Australia, Egypt, China, Peru, Argentina, and Brazil [1,2]. Each of these geographic regions places unique environmental constraints on the cultivars that can succeed there. Therefore, pecan producers have to select regionally and horticulturally-adapted cultivars and develop the required technologies for all steps of the production chain, from seed selection for nursery to nut harvesting for the industry. Since high quality seedlings are essential to establishing healthy and productive orchards, studies on new technologies for pecan cultivation in subtropical regions have mostly focused on seedling production [3,4].

Seed sizing is a single technique suggested to be an efficient indicator positively correlated with seed physiological quality and plantlet vigor [5,6]. Adams and Thielges [7] reported a weak positive correlation ($r = 0.21$) between seed size and seedling height in wild *C. illinoinensis*. In addition, these authors also reported absence of influence of seed weight on germination or seedling vigor in improved North American cultivars of pecan. Although climatic conditions, such as water availability, may interfere in this correlation [5], no study on this topic has been performed for pecan cultivars cultivated under subtropical conditions.

Given the scarce and inconclusive results concerning the correlation between seed mass and plantlet development in pecan, and the need for regionally adapted cultivars, we conducted a field experiment to examine the correlation between seed mass, plantlet emergence and seedling development of this species under subtropical climatic conditions.

Since the introduction of the pecan in Southern Brazil, farmers have searched for plants more adapted to the climatic conditions of the area, selecting cultivars that allowed the establishment of more healthy and productive orchards. Unpublished studies of our group suggest that the pecan cultivar *Importada* is one of the most promising for the subtropical climatic conditions of Southern Brazil. Therefore, this study evaluated the correlation between seed mass, seed emergence and plantlet vigor in this cultivar, towards producing healthy seedlings for orchard establishment.

2. Materials and Methods

2.1. Seeds Sampling and Classification

Seeds of the pecan cultivar *Importada* were collected from 10 mother-trees in Southern Brazil (28°58′18″ S, 52°00′30″ W; grown in a humid subtropical climate without a dry season, a *Cfa* climate according to the Köppen climate classification), generating a lot of 324 seeds. To overcome dormancy, seeds were maintained within wood boxes in moist sand at 4 °C ± 0.1 °C for 90 days, following Poletto and collaborators [4]. After this period, seed mass was determined using a precision balance (0.001 g) and seeds were sorted into seven classes according to their mass: Class I: <4.9 g (21 seeds); Class II: 5.0–5.9 g (45 seeds); Class III: 6.0–6.9 g (60 seeds); Class IV: 7.0–7.9 g (54 seeds); Class V: 8.0–8.9 g (57 seeds); Class VI: 9.0–9.9 g (54 seeds) and Class VII: >10.0 g (33 seeds).

2.2. Seeds Sowing and Data Collection

Seeds were sown manually by placing each seed in a 3-cm-deep hole, with a distance of 50 cm between rows and 20 cm between holes. Following the Brazilian System of Soil Classification [8], the soil of the region is Typic Argiudoll. Irrigation was performed through a drip system and weeds were manually controlled. The experiment was conducted using a completely randomized design consisting of 7 to 20 repetitions of three seeds for each seed mass class. The number of repetitions varied across seed mass classes according to the amount of seeds available in each category. Plantlet height (cm), stem diameter (mm), number of leaves and emergence rate ((total of emerged seedlings/total of sown seeds) × 100) were evaluated 16 weeks after sowing.

2.3. Statistical Analysis

Prior to statistical analysis, data on plantlet emergence were transformed for normalization employing $\sqrt{(x + 0.5)}$. The relationship between seed mass and each measured variable (plantlet height, stem diameter, number of leaves and emergence rate) was evaluated by regression analysis. Differences among seed mass classes were evaluated through the Kruskal–Wallis test, and pairwise comparisons were performed using the Mann–Whitney test with Bonferroni correction. Statistical analyses were implemented using the software PAST 3.02 [8] and Free Statistic Software v1.2.1 [9].

3. Results

3.1. Measures of Plantlet Vigor and Seedling Emergence

The differences in the measured traits between the smallest seed mass category (<4.9 g) and the largest seed mass category (>10.0 g) was 1.35-fold for plantlet height, 1.47-fold for stem diameter, 1.16-fold for leaf number, and 1.85-fold for emergence rate. The mean plantlet height ranged from 24.7 cm (Class I) to 33.4 cm (Class VII), the mean stem diameter ranged from 4.6 mm (Class I) to 6.7 mm (Class VI), and the mean number of leaves ranged from 11.4 in class I to 13.5 in class VI (Figure 1A–C). The mean emergence rate ranged from 52.4% in class I to 98.1% in class VI. Only one replicate in

Class VI and one in Class VII failed to reach 100% emergence, while none of the replications in Class I reached 100% emergence (Figure 1D).

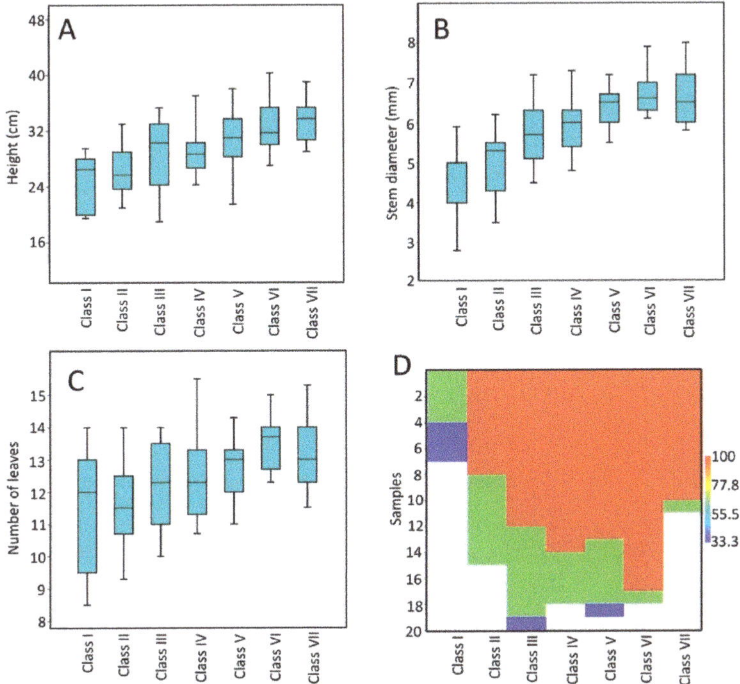

Figure 1. Boxplot diagrams showing the median (bar), 50% confidence intervals (box) and standard deviation (whiskers) for (**A**) plantlet height; (**B**) stem diameter; (**C**) number of leaves for each seed mass class; and (**D**) percentage of plantlet emergence determined as the percentage of germinated seeds across replicates for each seed mass class. Note that the number of replicates is different in each class (see text for details).

3.2. Correlation between Seed Mass, Seedling Emergence and Plantlet Vigor

A significant positive relationship between seed mass and emergence rate, plantlet height, stem diameter and number of leaves was observed in the regression analyses. Although the correlation between seed mass and the evaluated traits was almost linear, the best fit determined through the coefficient of determination (R^2) was observed using a polynomial model for all correlations (Figure 2).

Accordingly, the Kruskal–Wallis test revealed a statistically significant difference ($P < 0.001$) among seed mass categories for all measured traits, and the pairwise comparisons revealed a clear tendency for significant differences between small seed mass classes and larger seed mass classes (Table 1). Clearly, smaller seeds (Class I) revealed the weakest results and the best results were observed in the largest Classes VI and VII for all traits. The difference between means of classes I and VII were statistically significant for all traits except number of leaves (Table 1). The same pattern occurred when comparing means of Classes I and VI, while means from Classes VI and VII did not differ from each other for any of the traits (Table 1).

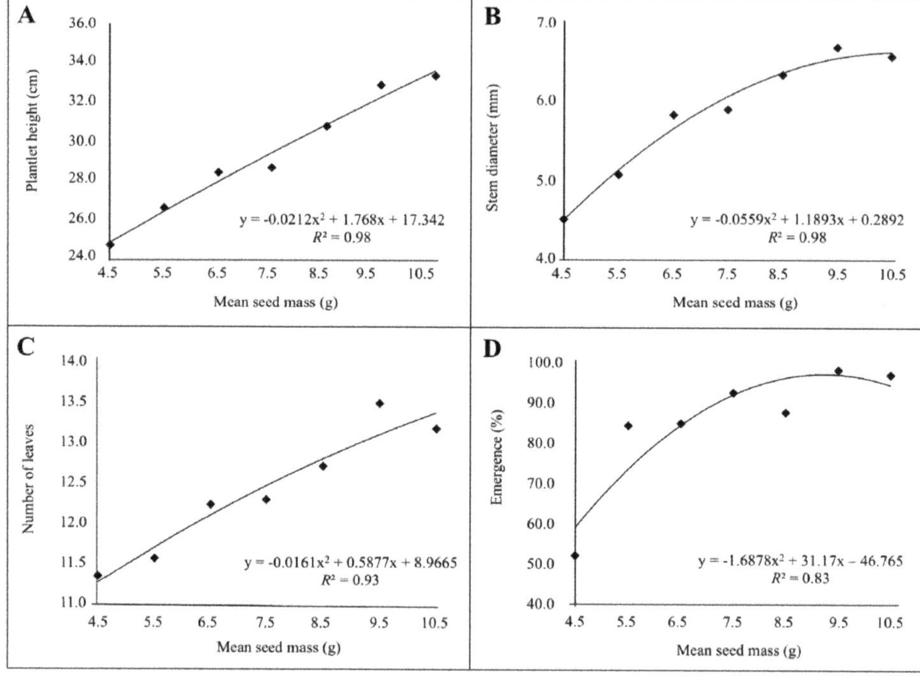

Figure 2. Regression analysis of seed mass of the pecan cv. Importada with means of (**A**) plantlet height; (**B**) stem diameter; (**C**) number of leaves and (**D**) emergence percentage. Linear regression equations and coefficients of determination (R^2) are presented for the relationships.

Table 1. *P*-values for the pairwise difference between classes for the evaluated traits, based on the Mann–Whitney test with Bonferroni correction. Bold numbers are statistically significant.

Class		I	II	III	IV	V	VI	Class	I	II	III	IV	V	VI
	II	1.00							1.00					
	III	1.00	1.00						0.23	0.45				
Height	IV	0.71	1.00	1.00				Stem	0.06	**0.04**	1.00			
	V	0.09	0.09	1.00	1.00			diameter	**0.01**	**0.00**	0.63	1.00		
	VI	**0.01**	**0.00**	0.55	**0.02**	1.00			**0.00**	**0.00**	**0.02**	**0.02**	1.00	
	VII	**0.02**	**0.00**	0.19	**0.03**	1.00	1.00		**0.02**	**0.00**	0.27	0.72	1.00	1.00
	II	1.00							0.13					
	III	1.00	1.00						**0.05**	1.00				
Number	IV	1.00	1.00	1.00				% of	**0.00**	1.00	1.00			
of	V	1.00	0.23	1.00	1.00			germination	**0.05**	1.00	1.00	1.00		
leaves	VI	0.15	**0.00**	0.03	0.06	0.19			**0.00**	0.15	0.29	1.00	0.94	
	VII	1.00	0.11	1.00	1.00	1.00	1.00		**0.01**	0.96	1.00	1.00	1.00	1.00

4. Discussion

During plant domestication, genotypes with larger seeds were selected for their contribution to increased grain yield, as well as for their positive effect on the quality of end-use products [10]. In our study, a significant correlation between seed size, seedling development (plantlet height and stem diameter) and emergence rate was observed for the pecan cultivar *Importada*. Seeds with the larger mass (Classes VI and VII, heavier than 9.0 g) produced larger seedlings and revealed higher emergence rates.

For commercial orchards, the heritability of these traits is an important factor. Thus, the genetic control of the variable determining the desired traits is vital. The control of seed size has been recognized as a complex trait and attributed to both genetic and epigenetic pathways. Based mainly on studies from cultivated cereal crops, new information is constantly added to the accumulated knowledge about this issue. For example, a study using high-throughput sequencing identified twenty-two microRNAs related to seed size in the shrub plant *Hippophae rhamnoides* [11], generating data which support future studies on genetic transformation.

The genetic control of seed size and yield in woody plants, on the other hand, has been investigated less. Recently, a homolog of *Auxin Response Factor 19* controlling seed size and seed yield was identified in *Jatropha curcas*, demonstrating the importance of the auxin pathway in seed size and yield control in woody plants [12]. Indeed, the transgenically-mediated overexpression of this gene in *J. curcas* and in *Arabidopsis thaliana* significantly increased seed size and seed yield in these plants [12].

Despite the limited knowledge about specific topics concerning the genetic control of seed size in woody plants, intraspecific variation in this trait is common in tree species. Evidence of a positive association between seed mass and mean time to germination was observed by Norden et al. [13] in an analysis of 1037 tropical tree species. Bispo et al. [6] reported direct interference of seed size in seedling development of *Anadenanthera colubrina*, with larger seeds yielding higher growth. Evaluating the capacity of seed fragments of *Eugenia pyriformis* to produce new seedlings, Pratavieira et al. [14] observed a significantly higher capacity by larger seeds. However, Moles and Westoby [15] found no relationship between seed mass and survival through the transition from seed to emerged seedlings in populations of 33 species growing under natural conditions. Jijeesh and Sudhakara [16] reported a lack of correlation between drupe size and the number of leaves, while the analysis of variance indicated that seedling height, collar diameter, leaf area, tap root length and number of lateral roots were significantly influenced by drupe size in *Tectona grandis* with larger drupes recording the highest seedling attributes.

Such positive correlations have been used as indicative of physiological quality, justifying the use of size classes for seed selection. In cork oak (*Quercus suber*), the largest seeds were correlated with plantlet survival in dry periods, suggesting the need for selection for this trait across sites experiencing drought [17]. The high positive correlation ($r > 0.81$) between seed mass, emergence rate and seedling development revealed in our study may be related to the selection procedures performed by the nurserymen. Selecting plus-trees for morphological characters leads to the 'hitchhiking' of some genes related to the desired traits. In cereal crops, seed size is a characteristic domestication attribute, and cultivated inbred lines and landraces usually have larger seeds than their wild progenitors. Accordingly, a signature of selection during domestication was observed in 63 out of 114 genes putatively related to seed size in *Sorghum bicolor* [10].

Different from the wild trees studied by Adams and Thielges [7], the pecan cultivar *Importada* has been selected for generations for traits such as nut size, tree development and adaptability to the subtropical climatic conditions of Southern Brazil. The seedlot of pecan used in this study may also express the effect of a genetic stratification of seed parents [18] because seeds were obtained from mother-trees cultivated with the aim of producing nuts for the food industry. Since these mother-trees were obtained from commercial nurseries, they were previously selected during seedling production.

5. Conclusions

Our study revealed a positive correlation between seed mass, seedling emergence and plantlet vigor, suggesting that seed mass can be used as a direct method for seed selection towards production of healthy pecan seedlings. However, in addition to superior emergence rate and seedling development, orchards should produce seeds with high breeding values and high gene diversity. Therefore, controlled studies—taking into consideration the recalcitrant nature of pecan seeds [19] and the influence of genetic stratification of seed parents on these traits—should be carried out. In addition, an increase in seed mass is usually associated with a decrease in the number of seeds that a plant can produce per unit canopy per year [20]. Thus, long-duration studies are recommended in order to evaluate this aspect. Furthermore, as suggested by Thompson and Conner [1], we are performing an update of the current genetic status of this crop with genomic characterization of the cultivars planted in subtropical regions, since breeding objectives have become more refined, and available methods of genetic plant improvement have expanded.

Author Contributions: Conceptualization, T.P. and I.P.; methodology, T.P.; formal analysis, T.P. and V.M.S.; resources, M.F.B.M.; writing—original draft preparation, T.P.; writing—review and editing, V.M.S.; supervision, M.F.B.M.; project administration, M.F.B.M.; funding acquisition, M.F.B.M. and V.M.S.

Funding: This research was funded by Conselho Nacional de Desenvolvimento Científico e Tecnológico (CNPq), Process 442995/2014-8.

Acknowledgments: Authors thank to Conselho Nacional de Desenvolvimento Científico e Tecnológico (CNPq) for fellowships to M.F.B.M. and V.M.S., and to Fundação de Amparo à Pesquisa do Estado do Rio Grande do Sul (FAPERGS) for the scholarship to T.P.

Conflicts of Interest: The authors declare no conflicts of interest.

References

1. Thompson, T.E.; Conner, P.J. Pecan. In *Fruit Breeding*; Badenes, M.L., Byrne, D.H., Eds.; Springer: New York, NY, USA, 2012; Volume 8, pp. 771–801, ISBN 978-1-4419-0762-2.
2. Zhang, R.; Peng, F.; Li, Y. Pecan production in China. *Sci. Hortic.* **2015**, *197*, 719–727. [CrossRef]
3. Cargnelutti Filho, A.; Poletto, T.; Muniz, M.F.B.; Baggiotto, C.; Poletto, I.; Fronza, D. Sampling design for height and diameter evaluation of pecan seedlings. *Ciênc. Rural* **2014**, *44*, 2151–2156. [CrossRef]
4. Poletto, I.; Muniz, M.F.B.; Poletto, T.; Stefenon, V.M.; Baggiotto, C.; Ceconi, D.E. Germination and development of pecan cultivar seedlings by seed stratification. *Pesqui. Agropecu. Bras.* **2015**, *50*, 1232–1235. [CrossRef]
5. Pereira, W.A.; Pereira, S.M.A.; Dias, D.C.F.S. Influence of seed size and water restriction on germination of soybean seeds and on early development of seedlings. *J. Seed Sci.* **2013**, *35*, 316–322. [CrossRef]
6. Bispo, J.S.; Costa, D.C.C.; Gomes, S.E.V.; Oliveira, G.M.; Matias, J.R.; Ribeiro, R.C.; Dantas, B.F. Size and vigor of *Anadenanthera colubrina* (Vell.) Brenan seeds harvested in Caatinga areas. *J. Seed Sci.* **2017**, *39*, 363–373. [CrossRef]
7. Adams, J.C.; Thielges, B.A. Seed sizes effects on first-and second-year pecan and hybrid Pecan growth. *Tree Planters' Notes* **1979**, *30*, 31–33.
8. Hammer, Ø.; Harper, D.T.; Ryan, P.D. Past: Paleontological Statistics Software Package for Education and Data Analysis. *Palaeontol. Electron.* **2001**, *4*, 1–4.
9. Wessa, P. Pearson Correlation (v1.0.13) in Free Statistics Software (v1.2.1), Office for Research Development and Education. 2017. Available online: https://www.wessa.net/rwasp_correlation.wasp/ (accessed on 5 March 2018).
10. Tao, Y.; Mace, E.S.; Tai, S.; Cruickshank, A.; Campbell, B.C.; Zhao, X.; Van Oosterom, E.J.; Godwin, I.D.; Botella, J.R.; Jordan, D.R. Whole-Genome Analysis of Candidate genes Associated with Seed Size and Weight in *Sorghum bicolor* Reveals Signatures of Artificial Selection and Insights into Parallel Domestication in Cereal Crops. *Front. Plant Sci.* **2017**, *8*, 1237. [CrossRef] [PubMed]
11. Ding, J.; Ruan, C.; Guan, Y.; Krishna, P. Identification of microRNAs involved in lipid biosynthesis and seed size in developing sea buckthorn seeds using high throughput sequencing. *Sci. Rep.* **2018**, *8*, 4022. [CrossRef] [PubMed]
12. Sun, Y.; Wang, C.; Wang, N.; Jiang, X.; Mao, H.; Zhu, C.; Wen, F.; Wang, F.; Lu, Z.; Yue, G.; et al. Manipulation of *Auxin Response Factor 19* affects seed size in the woody perennial *Jatropha curcas*. *Sci. Rep.* **2017**, *7*, 40844. [CrossRef] [PubMed]
13. Norden, N.; Daws, M.I.; Antoine, C.; Gonzales, M.A.; Garwood, N.C.; Chave, J. The relationship between seed mass and mean time to germination for 1037 tree species across five tropical forests. *Funct. Ecol.* **2009**, *23*, 203–210. [CrossRef]
14. Prataviera, J.S.; Lamarca, E.V.; Teixeira, C.C.; Barbedo, C.J. The germination success of the cut seeds of *Eugenia pyriformis* depends on their size and origin. *J. Seed Sci.* **2017**, *37*, 47–54. [CrossRef]
15. Moles, A.T.; Westoby, M. Seedling survival and seed size: A synthesis of the literature. *J. Ecol.* **2004**, *92*, 372–383. [CrossRef]

16. Jijeesh, C.M.; Sudhakara, K. Larger drupe size and earlier geminants for better seedling attributes of teak (*Tectona grandis* Linn. f.). *Ann. For. Res.* **2013**, *56*, 307–316.
17. Ramírez-Valiente, J.A.; Valladares, F.; Gil, L.; Aranda, G. Population differences in juvenile survival under increasing drought are mediated by seed size in cork oak (*Quercus suber* L.). *For. Ecol. Manag.* **2009**, *257*, 1676–1683. [CrossRef]
18. Chaisurisri, K.; Edwards, D.G.W.; El-Kassaby, Y.A. Effects of seed size on seedling attributes in Sitka spruce. *New For.* **1994**, *8*, 81–87. [CrossRef]
19. Dalkiliç, Z. Effects of drying on germination rate of pecan seeds. *J. Food Agric. Environ.* **2013**, *11*, 879–882. [CrossRef]
20. Henery, M.L.; Westoby, M. Seed mass and seed nutrient content as predictors of seed output variation between species. *Oikos* **2001**, *92*, 479–490. [CrossRef]

MDPI
St. Alban-Anlage 66
4052 Basel
Switzerland
Tel. +41 61 683 77 34
Fax +41 61 302 89 18
www.mdpi.com

Horticulturae Editorial Office
E-mail: horticulturae@mdpi.com
www.mdpi.com/journal/horticulturae